Educators' Learning from Lesson Study

Offering voices from the field – the first of its kind outside of Japan – this guide to teaching and learning elementary mathematics highlights real case examples from teachers and educators who share what they have learned through Lesson Study.

The teachers' reports provide vivid examples of new insights and ideas about mathematics, about pedagogy and lesson design, about student learning, and about professional collaboration gained through Lesson Study. Each report includes an abbreviated plan of the specific research lesson that led to the new insights, which readers can draw from to replicate the powerful learning in their own community. The case examples of this book are from Lesson Study in mathematics, elementary to lower secondary grade levels, focused on what teachers and educators have learned about improving mathematics teaching and learning; but many ideas from each report can be applied to other subjects and different grade levels.

This unique book will be an excellent resource for mathematics teachers in training and practice who seek to improve mathematics teaching and learning in their own and others' classrooms, including researchers and school administrators who lead professional development.

Akihiko Takahashi is Associate Professor of Mathematics Education at DePaul University. He received his Ph.D. from the University of Illinois at Urbana-Champaign and has published over 80 journal articles and book chapters in English and Japanese. He has given over 50 presentations and keynotes at conferences and workshops in 18 countries.

Thomas McDougal is Founder & Executive Director of Lesson Study Alliance in Chicago. He has worked previously as a high school math teacher and as an elementary (grades K–8) mathematics specialist in the Chicago Public Schools. In 2016, he co-authored the paper, "Collaborative Lesson Research: Maximizing the Impact of Lesson Study."

Shelley Friedkin is Senior Research Associate for the Lesson Study Group at Mills College. Her research interests include teachers' learning from practice and practice-based artifacts, and developing school-wide Lesson Study.

Tad Watanabe is Professor of Mathematics Education and an associate chair of the Department of Mathematics at Kennesaw State University. His research interests include teaching and learning of multiplicative concepts and Lesson Study.

WALS-Routledge Lesson Study Series
Series editors: Christine Kim-Eng Lee, Catherine Lewis, Kiyomi Akita, and Keith Wood

This series aims to provide opportunities for researchers and practitioners in Lesson Study to share their work beyond the boundaries of their countries to an international audience. Lesson Study is increasingly popular as a tool for improving the quality of education and schools around the world. Many countries are adapting and contextualizing Japanese Lesson Study to their own needs in response to educational and curriculum reforms cognizant that what matters most is what happens in classrooms and its impact on teachers and students. As Lesson Study originates from Japan, there is also a need for English Language readers around the world to understand more deeply the underlying philosophies, policies and practices of Japanese Lesson Study in the cultural contexts of their schools and classrooms. As well as original works in English from leading figures in Lesson Study, this series will also make available outstanding Lesson Study publications originally written in Japanese but extended and revised for an English audience.

Educators' Learning from Lesson Study
Mathematics for Ages 5–13

**Edited by Akihiko Takahashi,
Thomas McDougal, Shelley Friedkin, and
Tad Watanabe**

Routledge
Taylor & Francis Group

LONDON AND NEW YORK

Cover image: © Getty Images

First published 2022
by Routledge
4 Park Square, Milton Park, Abingdon, Oxon OX14 4RN

and by Routledge
605 Third Avenue, New York, NY 10158

Routledge is an imprint of the Taylor & Francis Group, an informa business

British Library Cataloguing-in-Publication Data
A catalogue record for this book is available from the British Library

Library of Congress Cataloging-in-Publication Data
Names: Takahashi, Akihiko, 1955- editor. | McDougal, Thomas, 1962- editor. | Friedkin, Shelley, 1965- editor. | Watanabe, Tad, 1959- editor.
Title: Educators' learning from lesson study : mathematics for ages 5-13 / edited by Akihiko Takahashi, Thomas McDougal, Shelley Friedkin, Tad Watanabe.
Description: Abingdon, Oxon ; New York, NY : Routledge, 2022. | Series: WALS-Routledge lesson study series | Includes bibliographical references and index.
Identifiers: LCCN 2021057871 (print) | LCCN 2021057872 (ebook) | ISBN 9781032138176 (hardback) | ISBN 9781032138169 (paperback) | ISBN 9781003230915 (ebook)
Subjects: LCSH: Mathematics—Study and teaching (Elementary) | Mathematics teachers—In-service training.
Classification: LCC QA135.6 .E33 2022 (print) | LCC QA135.6 (ebook) | DDC 372.7/044—dc23/eng20220314
LC record available at https://lccn.loc.gov/2021057871
LC ebook record available at https://lccn.loc.gov/2021057872

ISBN: 978-1-032-13817-6 (hbk)
ISBN: 978-1-032-13816-9 (pbk)
ISBN: 978-1-003-23091-5 (ebk)

DOI: 10.4324/9781003230915

Typeset in Times New Roman
by Apex CoVantage, LLC

"The editors of this volume are the top practitioners and supporters of Lesson Study in the US, and possibly the world. This volume shows what teachers can learn when engaging deeply in Lesson Study and supported by masters."

– **Alan Schoenfeld**, *Elizabeth and Edward Conner Professor of Education, University of California, Berkeley*

"From teachers themselves, learn the power of Lesson Study for teaching and learning mathematics. As teacher teams practice essential elements of Lesson Study, their own knowledge of mathematics grows and they find multiple strategies that help or hinder individual students. Learn what is necessary for this remarkable form of professional development to forge exciting learning communities in schools that bring out the best in children. This book is a testament to why Lesson Study is worth the effort!"

– **Alice J. Gill**, *Retired Senior Associate Director, AFT Educational Issues Department*

"Lesson Study works at large scales of implementation: it is used in virtually every elementary school in Japan and widespread across many Asian countries. It works to build teachers' pedagogical content knowledge. They learn more math month after month by studying students' mathematical thinking shoulder to shoulder with colleagues. Lesson Study focuses on the mathematics in student thinking in response to teaching. It is the best way to improve implementation of your program. This book gives you the wisdom of worldwide leaders and the experience of American practitioners of lesson study."

– **Phil Daro**, *Director of Mathematics, Strategic Education Research Partnership (SERP)*

"A novel collection curated from in-depth work by classroom practitioners, guided by experts in the field of lesson study, this book is a must-read for educators interested in using lesson study to improve teacher knowledge and learning process in mathematics. The excellent juxtaposition of research and practice makes this book an excellent resource to those already practicing lesson study as well as to those looking for inspiration in starting a learning community among colleagues."

– **Yeap Ban Har**, *Pathlight School, Singapore*

For all who work so hard through Lesson Study to benefit their students

Contents

x *Contents*

Figures

Tables

Foreword

Milestones and miles to go: the power of educators' Lesson Study reports

Readers have a treat ahead. This volume is both an important milestone and a valuable resource. It is a milestone because we hear directly from classroom teachers who are using Lesson Study to solve their pressing problems of practice. For example, we learn how to help students connect their conceptual understanding of multi-digit multiplication and division to the procedural algorithms students must master. It is a milestone because the book's teacher-authors place problem-solving at the heart of their mathematics instruction, a goal that has broadly eluded U.S. education for more than 40 years (NCTM, 1980; Banilower et al., 2018). It is a milestone because, in the classrooms of these educators, *students* build each new mathematical concept and procedure in the curriculum, first by working independently on a carefully conceived problem, and then by participating in whole-class discussion that builds the new mathematics from careful examination of student work. This instructional approach, Teaching Through Problem-solving, was highlighted by *The Teaching Gap* (Stigler & Hiebert, 1999), which credited it for Japan's outstanding performance on international mathematics assessments. Unfortunately, *The Teaching Gap*'s insight that "teaching is a cultural practice" is often used to argue that Teaching Through Problem-solving is unlikely to succeed outside Japan. If you share that view, this volume will change your mind.

This book is a valuable resource because educators explain *how* they use Lesson Study to build their own content and instructional knowledge and to support a dramatic transformation from "teaching as telling" to Teaching Through Problem-solving (TTP). Brief, powerful chapters explain how teacher teams conduct content study (*kyouzai kenkyuu*) for many mathematical topics (such as addition-subtraction word problems, decimals, and fraction addition). Teachers recount what they have learned from observing students during live research lessons, taking part in a post-lesson discussion, and hearing final comments from an outside mathematics educator who has observed the lesson – experiences that are remarkably *un*common in much U.S. professional learning. Teachers capture the satisfaction of working closely with colleagues to identify and work on a challenge they all care deeply about, such as nurturing an understanding of place value that will serve students well in future grades. Chapters capture the delight teachers feel as they inch closer toward being the kind of teacher they want to be, with newfound support from TTP instructional strategies – such as reflective mathematics journals, planned board use, and student-led whole-class discussion – that powerfully elevate students' voices and ideas. Chapters capture the power of doing Lesson Study in teams throughout a school, united by a school-wide vision of student-centered discourse embraced by all your colleagues. And I would be remiss to gloss over the time and hard work these educators also portray – the meetings that lasted late into the dark of a winter afternoon in Chicago, the many hours of prep time invested by a second-year teacher who agreed to open up her research lesson to a hundred observers as part of a Lesson Study conference.

This book illuminates how Lesson Study – a professional learning approach – and Teaching Through Problem-solving – an approach to mathematics instruction – work in synergy. Lesson Study provides a structure for teachers to collaborate and learn – from study of curriculum and research, from co-planning a unit and research lesson together, and from observing and reflecting on student responses to the research lesson. TTP makes student thinking visible during the research lesson, through instructional routines such as board use, journals, and discussion (LSGAMC, n.d.-a). Because TTP places strong demands on teachers' content knowledge (to grasp the mathematics in student thinking), it needs some Lesson Study-like process to support teachers' study of content and anticipation of student thinking. Although Lesson Study can be used to study virtually any content area or instructional approach, it is most successful when student thinking is visible, as occurs during TTP lessons. We learn how San Francisco's Teacher Leader Fellowship has extended Lesson Study to support new initiatives, such as Problem-Based Learning in science, after finding success using Lesson Study with TTP. This volume allows a close examination of the synergy between TTP and Lesson Study that will help readers from other disciplines and instructional approaches imagine how to apply Lesson Study to their domain of interest.

So, if these chapters are so powerful, why does my title include "miles to go"? This teacher-written volume provides powerful examples that will enable other teachers to transform mathematics teaching and learning and to nurture students who experience the power of their own mathematical thinking. As a young student at one of the whole-school Lesson Study sites said, "The teacher doesn't show the math. The kids show our math. We learn better when we show our ways." So it should not be surprising that school-wide Lesson Study sites using TTP have shown dramatic improvements in students' mathematics outcomes, sharply reversing patterns of inequity typically associated with poverty, race, ethnicity, and language status (see the chapter by Akihiko Takahashi and also LSGAMC, n.d.-b). But district administrators, policymakers, and researchers must join teachers in the remaining miles of the journey; they must notice and elevate the voices of these site-based educators, and use their insights to redesign district systems that support teacher learning. Districts play a crucial role in the elevation or demise of successful school-wide Lesson Study, as some sobering prior examples reveal (Perry & Lewis, 2010).

How should readers use this book? If you are a teacher of mathematics who is new to Lesson Study, consider starting with Berenice Heinlein's case study, "Designing Lessons Students Lead," to learn how observing a research lesson might start your Lesson Study journey. If you are already involved in Lesson Study, you will find many chapters to spark reflection and experimentation on your team's content study, instructional design, and processes for learning from lesson observation and post-lesson discussion. If you are a researcher, look for ways that these accounts expand and challenge your ideas about teachers' and students' learning within Lesson Study. If you are a coach or administrator, notice that these accounts capture important keys to equitable student outcomes, in their emphasis on anticipating and designing for student thinking and observing how lessons actually play out in classrooms. Notice the thoughtfulness and agency in these teacher-written accounts, and consider whether your current structures for teacher and student learning support (or perhaps unwittingly undermine?) the powerful work of teachers like these. If you are a member of the broad international WALS community, let's think together about how voices like those captured in this volume, from practicing educators in the rest of the world, can shape future WALS conferences and publications.

My deepest thanks to the authors and editors of this volume. I hope it will inspire many practicing educators to share their work in future volumes that extend across disciplines, grade levels, regions of the world, and instructional approaches.

– Catherine Lewis, Mills College, Oakland CA

This material is based upon work supported by the Bill and Melinda Gates Foundation; the Institute of Education Sciences, U.S. Department of Education (through Grants R305A150043, R305A110491 and R305A070237 to Mills College); and the National Science Foundation (through Grant No. 0207259). The opinions expressed are those of the author and do not necessarily reflect the views of the funders.

References

Banilower, E. R., Smith, P. S., Malzahn, K. A., Plumley, C. L., Gordon, E. M., & Hayes, M. L. (2018). *Report of the 2018 NSSME+*. Horizon Research, Inc.

Lesson Study Group at Mills College. (n.d.-a). *Teaching through problem-solving*. https://lessonresearch.net/teaching-problem-solving/overview

Lesson Study Group at Mills College. (n.d.-b). *School-wide Lesson Study: John Muir Elementary School*. https://lessonresearch.net/john-muir-elementary/overview

NCTM (National Council of Teachers of Mathematics). (1980). *An agenda for action*.

Perry, R., & Lewis, C. (2010). Building demand for research through lesson study. In C. E. Coburn & M. K. Stein (Eds.), *Research and practice in education: Building alliances, bridging the divide* (pp. 131–145). Rowman & Littlefield Publishers, Inc.

Stigler, J. W., & Hiebert, J. (1999). *The teaching gap: Best ideas from the world's teachers for improving education in the classroom*. Summit Books.

Preface

Starting in the late 1990s, numerous teachers at schools outside Japan have tried to use Lesson Study to improve teaching and learning. In the US, the effort was primarily led by researchers and professional development specialists. Although researchers reported some benefits of using Lesson Study to improve teaching, Lesson Study almost never took hold as a sustained approach to teacher growth, and significant impacts on student learning were rarely documented.

After years of trial and error, some Lesson Study projects outside Japan gradually impacted both teacher and student learning. Some projects tried to adopt Lesson Study commonly seen in schools in Japan faithfully, and some projects adapted Lesson Study to their school systems and teacher needs (e.g., Hart, Alston, & Murata, 2011; Huang, Takahashi, & da Ponte, 2019; Quaresma et al., 2018). Based on these experiences, researchers identified critical conditions and strategies for impactful implementation of Lesson Study schools outside Japan (e.g., Lewis & Hurd, 2011; Lewis, Perry, Hurd, & O'Connell, 2006; Takahashi & McDougal, 2016). These reports discussed how teachers gained new insights for improving teaching and learning, and how their classrooms became a place where students actively engage in studying mathematics.

Even though many teachers have now been part of Lesson Study projects in various countries, few articles and books have been published by those teachers. In Japan, on the other hand, it is more common for articles and books to be authored by the teachers who conduct Lesson Study than by researchers who observe them. These books and reports are popular: Japanese teachers often refer to these reports and books when they prepare their everyday lessons and research lessons. We, the editors, believe that the time is overdue to invite the teachers involved in Lesson Study to share what they have learned with the world.

This book, written by 18 US teachers, educators, and researchers who have been practicing Lesson Study, reports what they learned from Lesson Study: about the contents that they teach; pedagogical ideas to enhance student learning; the nature of students' learning, including challenges that students encounter; and the benefits of teacher collaboration. In the first chapter, three editors summarize how Lesson Study can help teachers learn and develop new teaching and learning ideas and insights. Chapters 2 through 5 contain specific case studies of educator learning through Lesson Study. In Chapter 2, teachers and educators report what they learned about the contents they teach. In Chapter 3, the cases focus on pedagogical ideas and lesson design. In Chapter 4, teachers share insights gained into student learning. Chapter 5 addresses teacher leadership and collaboration. The last chapter, Chapter 6, provides a summary and proposes ideas to maximize teachers' learning in Lesson Study.

All the chapters are intended for practitioners to use as a resource for Lesson Study as well as for planning daily lessons.

We hope this collection of Lesson Study reports written by teachers, teacher leaders, and school administrators who care about student learning encourages more teachers and schools worldwide to try Lesson Study and to share what they learn.

Akihiko Takahashi
Thomas McDougal
Shelley Friedkin
Tad Watanabe

References

Hart, L. C., Alston, A., & Murata, A. (Eds.). (2011). *Lesson Study research and practice in mathematics education*. Springer.

Huang, R., Takahashi, A., & da Ponte, J. (Eds.). (2019). *Theory and practice of Lesson Study in mathematics: An international perspective*. Springer.

Lewis, C., & Hurd, J. (2011). *Lesson Study step by step: How teacher learning communities improve instruction*. Heinemann.

Lewis, C., Perry, R., Hurd, J., & O'Connell, M. P. (2006). Lesson Study comes of age in north America. *Phi Delta Kappan, 88*(4), 273–281.

Quaresma, M., Winsløw, C., Clivaz, S., Ponte, J. P., Shúilleabháin, A. N., & Takahashi, A. (Eds.). (2018). *Mathematics Lesson Study around the world*. Springer International Publishing.

Takahashi, A., & McDougal, T. (2016). Collaborative lesson research: Maximizing the impact of lesson study. *ZDM Mathematics Education, 48*(4), 513–526. doi:10.1007/s11858-015-0752-x

Acknowledgements

This book, written by 18 US teachers, educators, and researchers, is a product of the tireless efforts of US teachers and school leaders to improve mathematics teaching and learning using Lesson Study. Because Lesson Study is still an unconventional approach to teacher learning in the US, none of the schools at which these educators worked initially had structures or resources in place to support Lesson Study. These pioneers of Lesson Study had to develop a plan to introduce this new approach, adapt it to be sustainable, and secure funding and resources.

We must start by giving thanks to those teachers who spent many hours developing the research lessons cited in this volume. Even if they did not write a case study for this volume, their work made this volume possible.

We also owe a deep gratitude to those school leaders who made significant sacrifices of time, energy, and resources to ensure that their teachers could engage in Lesson Study, and who generously opened their schools for public lessons so that others could benefit. In Oakland: Leroy Gaines at Acorn Woodland Elementary School. In San Francisco: Shawn Mansager and Sara E. Liebert at John Muir Elementary School. In Los Gatos: Kit Bragg, formerly at Daves Elementary School. In Chicago: Barton Dassinger at Chavez Multicultural Academic Center; Lorianne Zaimi at Helen Peirce Elementary; Seth Lavin and Erendira Alcántara at Brentano Math and Science Academy; Mariel Laureano, Amber Richard, and Andrew Friesema at Prieto Math and Science Academy. Every spring so far since 2012, Prieto has hosted the Chicago Lesson Study Conference, which was the setting for several of the research lessons described in this volume. Laureano, Richard, Friesema and other staff at Prieto deserve extra recognition and thanks. Not only did they let us take over their gym for two days each year (thank you Kevin Drumm!), but they provided invaluable behind-the-scenes logistical support.

We also send gratitude to Courtney Ortega and Robin Lovell from the math department at Oakland Unified School District and to Nora Houseman and her team from San Francisco Unified School District's Office of Professional Learning and Leadership.

Vivian Mihalakis, the program officer for the school-wide Lesson Study funded by the Gates Foundation, was a supportive and thoughtful sounding board for the work in Chicago, San Francisco, and Oakland.

Finally, we give thanks to the generous public and private organizations whose financial support helped make the Lesson Study work described in this volume happen. Critical support came from the Bill & Melinda Gates Foundation; the Institute of Education Sciences, US Department of Education (through Grants R305A150043, R305A110491, and R305A070237 to Mills College); the National Science Foundation (through Grant No. 0207259); the McDougal Family Foundation; and Project IMPULS at Tokyo Gakugei University.

Contributors

Aaron Bingea	Brentano Math and Science Academy
Brigid Brown	Oakland Unified School District (formerly)
Shelley Friedkin	Mills College (formerly)
Berenice Heinlein	Helen C. Peirce School of International Studies
Nora Houseman	San Francisco Unified School District
Alexandra Johansen Laughlin	Dr. Jorge Prieto Math and Science Academy (formerly)
Cassie Kornblau	Brentano Math and Science Academy
Kari Laux	Citizens of the World Charter School
Joshua Lerner	Helen C. Peirce School of International Studies
Sara E. Liebert	John Muir Elementary School
Thomas McDougal	Lesson Study Alliance
Jana Morse	Independent Consultant
Aubrey E. Perlee	Dr. Jorge Prieto Math and Science Academy
Rebecca Reddicliffe	Brentano Math and Science Academy
Lindsay Singer Kalt	Cesar Chavez Multicultural Academic Center (formerly)
Meghan Smith	Latin School of Chicago
Akihiko Takahashi	DePaul University/Lesson Study Alliance
Tad Watanabe	Kennesaw State University

I How we expect Lesson Study to contribute to the quality of teaching and learning

DOI: 10.4324/9781003230915-1

Teacher as life-long learner

Shelley Friedkin[1]

Lesson Study is a deep dive that has touched on my values as a teacher and my knowledge foundation. When we reflect at the end of a Lesson Study cycle, it shows me how to carry my learning beyond the cycle and impacts how I show up daily for my students in the classroom.

– Brigid Brown, 3rd grade teacher, Oakland, CA

When the first English-language publications on Lesson Study appeared in 1998–99, the historical context of teacher learning and mathematics instructional improvement in the U.S. provided the backdrop to Lesson Study's implementation. An influential report, *A Nation at Risk: The Imperative for Educational Reform* (U.S. National Commission on Excellence in Education, 1983) had called for extensive reform of the U.S. education system, highlighting the need for better teaching. This report was the first to raise the notion of teacher learning and teacher development as an important focus to improve U.S. education. Underlying this new idea was the teaching profession's historic culture valuing teachers as all-knowing, self-sufficient, and certain (Lortie, 1975). As such, reimaging teachers as learners lay ahead for Lesson Study, as did the undoing of centuries of occupational socialization valuing teachers as already competent and discouraging questioning, sharing classroom failures, or openly admitting frustration (Cochran-Smith & Lytle, 1993).

By the turn of the 21st century, U.S. national organizations and state agencies guidelines and frameworks started to put a spotlight on teachers and teaching practice (California State Department of Education, 1992; National Council of Teachers of Mathematics, 1989, 2001). These documents articulated the need for a shift in mathematics instruction from the teacher being the dispenser of knowledge and teaching as telling, to the teacher as facilitator and teaching for student understanding. New and ambitious visions about teaching and learning started to emerge based on a better understanding in the field about how students learn (Fuson, Kalchman, & Bransford, 2005; National Research Council, 2000). Researchers believed that students should engage in cooperative work, discuss, question, and problem solve. Evidence connecting such practices and students' acquisition of mathematical understanding revealed that when instruction focused on reasoning and sense making, rather than on memorization and procedures, students' capacity to solve complex mathematical problems increased (Stein & Lane, 1996).

DOI: 10.4324/9781003230915-2

Help in meeting higher expectations for mathematics teaching and learning initially arrived in the form of new curricular materials and these resources provided more challenging instructional tasks (Smith, Stein, & Silver, 2005). Nonetheless, many teachers remained uncertain how to change their teaching to accommodate the reform ideas of teaching as facilitation (Heaton, 2000; Silver, Ghousseini, Gosen, Charalambous, & Font Strawhun, 2005; Wilson, 2003). For example, how do teachers support students to solve challenging mathematical problems without taking over the process of thinking for them, and without eliminating the challenge (National Council of Teachers of Mathematics, 2000)?

To support teachers as they explored ways of facilitating student learning, professional development shifted from short intense subject matter presentations to be practice-based and on-going (Darling-Hammond, 1998; Lord, 1994; Schon, 1987). Practiced-based professional learning emerged and held the intent to be teacher-driven, collaborative, and reflective, and provide opportunities for teachers to solve their own problems of practice (Annenberg Institute for School Reform, 2005; Dufour, 2001; Lieberman & Wood, 2002; Snow-Gerono, 2005).

Lesson Study and teacher learning

The book *The Teaching Gap* (Stigler & Hiebert, 1999) brought broad public attention to previous scholarship on Lesson Study (Lewis & Tsuchida, 1998; Yoshida, 1999) and speculated the higher student achievement in mathematics in Japan could be attributed to Lesson Study. *The Teaching Gap* pushed for Lesson Study to be pursued in the U.S. because it laid out "a clear model for teacher learning and a clear set of principles or hypotheses about how teachers learn" (p. 150).

Lesson Study's premise is that teachers inquire, examine, and gain a clearer understanding about the issues of teaching and learning by observing and gathering data from actual classroom lessons (Lewis, Perry, & Hurd, 2004). The Lesson Study cycle involves teachers working collaboratively to study a topic, plan a lesson, teach a lesson, collect student data, reflect and debrief their practice (Lewis, 2002). This process of gathering evidence about student thinking and drawing inferences informs teachers about students' challenges or progress toward learning goals and subsequently supports decisions about appropriate instructional action (Bryant & Driscoll, 1998; Cochran- Smith, 2006; Franke & Kazemi, 2001; Pellegrino, Chudowksy, & Glaser, 2001).

An early challenge identified for teachers doing Lesson Study was taking on the role of a researcher (Fernandez, Cannon, & Chokshi, 2003; Lewis, Perry, & Murata, 2006). Fernandez et al. (2003) documented how Japanese teachers mentored American teachers in Lesson Study and observed "the Japanese teachers continually encouraging the American teachers to see themselves as researchers conducting empirical examination, organized around asking questions about practice and designing classrooms experiments to explore these questions" (p. 173). The problem appeared to be that U.S. teachers did not have experience carrying out systematic inquiry cycles and did not see research on practice as part of their role as teachers. Quite likely, this is because research and inquiry into teaching has

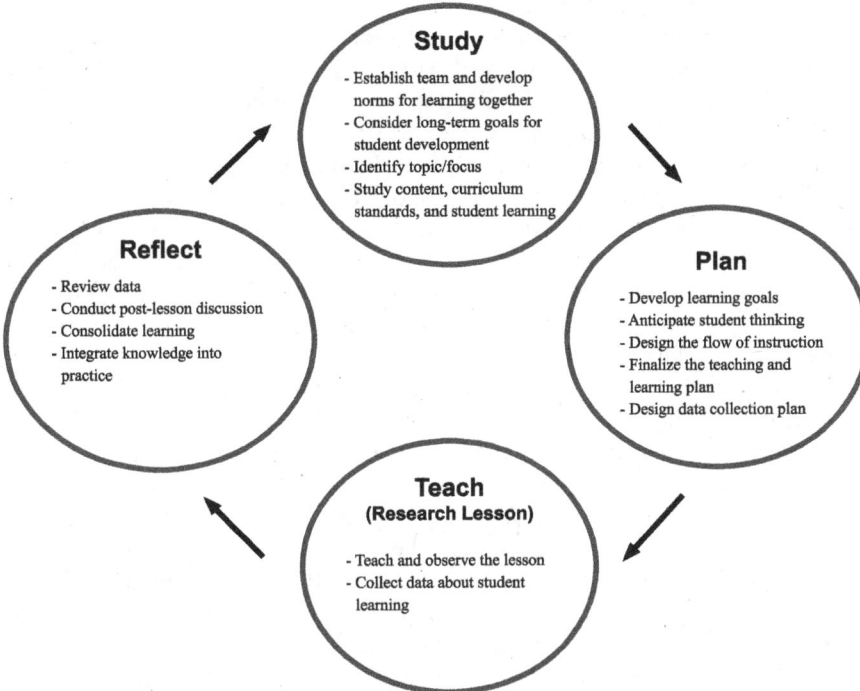

Figure 1.1 Lesson Study cycle

Source: Adapted from Lesson Study Research Group at Mills College LSRG, 2021a

been the domain of university-based scholars. Cochran-Smith & Lytle describe how teachers are often the "object of study", rather than participants in the research process. Teachers' classrooms were typically sites for data collection by outsiders rather than sites where they collect and analyze data themselves. They continue,

> Lack of significant teacher participation in codifying what we know about teaching, identifying research agendas, and creating new knowledge is problematic. Those who have daily access, extensive expertise, and a clear stake in improving classroom practice have no formal ways for their knowledge of classroom teaching and learning to become part of the literature on teaching.
> (Cochran-Smith & Lytle, 1993, p. 5)

A move toward recognizing that teachers hold significant and unique forms of knowledge that contribute to research about teaching and learning is underway. In presenting this book, we take an important step toward exemplifying the unique forms of teacher knowledge that contribute to not only research efforts but toward identifying, framing, articulating, and advancing the field's knowledge about teaching and learning mathematics.

U.S. Lesson Study is now in its 22nd year (if we agree that Lesson Study emerged in the U.S. in 1999). During this time, a depth of information from around the world has been published describing the kinds of knowledge that teachers gain during Lesson Study. Huang and Shimizu (2016), in their comprehensive review of mathematics Lesson Study, report on teacher learning and include knowledge domains such as content knowledge, pedagogical knowledge, pedagogical content knowledge, and knowledge about student thinking. Researchers exploring social-cognitive shifts during Lesson Study report on teachers' sense of efficacy, beliefs, agency, motivation, and curiosity (Chong & Kong, 2012; Dudley, 2013; Lewis, Fischman, Riggs, & Wasserman, 2013; Lewis & Perry, 2015, Lewis, Friedkin, Emerson, Henn, & Goldsmith, 2019; Puchner & Taylor, 2006; Schipper, 2019; Sibbald, 2009). Another focus within the Lesson Study research community is examination of teachers' collaboration, where teachers with differing levels of experience and expertise discuss, negotiate, and develop a shared vision about teaching and learning. From this exploration, the collaborative structures and supports integral to building joint ownership and responsibility during the Lesson Study process are now better understood. For example, collaborative structures such as norm setting, facilitation of learning discussions, post-meeting reflection prompts, and protocols for observing the research lesson and collecting data are well documented and shared freely (see *www.lessonresearch.net*, *www.lsalliance.org*).

A shared framework describing how teachers learn during collaborative practice-based professional learning environments, such as Lesson Study, is still lacking in the field (Boylan, Coldwell, Maxwell, & Jordan, 2018). The Lesson Study Research Group at Mills College (LSRG, 2021b), in an attempt to address this gap and start a dialogue, share several models of teacher learning at their website (*www.lessonresearch.net/teacher-learning*). One framework, in particular, focuses on four dimensions of teacher learning that occur during the Lesson Study process; it includes knowledge, beliefs and dispositions, agency, and collegial learning culture (LSRG, 2021c). Collegial learning culture incorporates both *collaborative structures*, such as those described previously (e.g. norm setting and discussion protocols), and *collegial relationships*. Collegial relationships capture the dynamic between Lesson Study team members that allows them to learn from one another.

While the lesson study phases (*study, plan, teach, reflect*), by design, support teacher research into classroom practice, and teachers' growth in the four dimensions allow teachers to improve their Lesson Study work, it is the underlying sensitives to the inquiry at hand, referred to here as the *inquiry stance*, that is considered most likely to predict the quality of teacher learning during Lesson Study. The inquiry stance works in symbiotic relationship with the Lesson Study phases and the four dimensions of teacher growth. The inquiry stance binds the Lesson Study team together, and it keeps the learning alive and relevant for all teachers on the team.

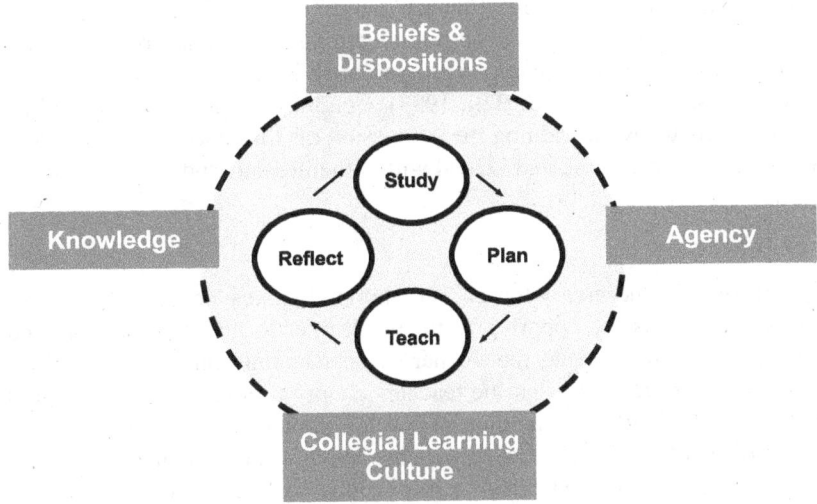

Figure 1.2 Four dimensions of teacher learning

Source: Lesson Study Research Group at Mills College LSRG, 2021c. Reproduced with permission

How the inquiry stance predicts the quality teacher learning during Lesson Study

During the Lesson Study process, several key conditions appear to support teachers' learning while also strengthening the inquiry stance. These conditions include access and opportunity, use of evidence, and the learners' disposition toward development and growth, or to be a life-long learner.

Access and opportunity

The foremost condition necessary for teacher learning to take place within the Lesson Study community is access to learning: the empowerment of each teacher to participate is dependent upon the access afforded within the community of practice – for example, the structures in place that enable the learner to "see, hear and participate in the work in question: its central tasks, tools and instruments, relevant categories and terms and lines of communication" (Little, 2003, p. 917). This access includes providing teachers the opportunity to frankly express their individual understanding and to ask authentic questions. If prior ways of knowing remain hidden constructions of personal identity (historical, political, moral, and cultural values, beliefs, and assumptions), then learning will remain shallow. According to Little (2003), an individual's or group's capacity for airing, acknowledging, and responding to differences and conflict correlate directly with the opportunity for sustained and deep consideration of teaching problems and possibilities. Such a culture of honest expression and reflection gives rise to perturbations and dissonance necessary for learning to take place by providing an opportunity for a genuine awareness of alternative ideas and openness to rethink one's own position (Chazan & Ball, 1999). Conditions that build trust and joint challenge or focus, while valuing the expression of differences, are necessary to keep learning alive, shared, and valued within the Lesson Study community.

Use of evidence

A second condition necessary for teacher learning to take place within the Lesson Study community is the opportunity for the learner to make connections: connecting learning by situating the learner's understanding through an artifact of research. Artifacts refer to what the teachers choose to focus on, to share or elevate to illustrate their understanding of classroom practice and student learning. Ball and Cohen (1999) advocate for teacher learning to be both situated within practice and to be grounded in the "real phenomena of practice. Such concreteness can be the beginning of serious intellectual work rather than a low-grade alternative, focused on more personal opinion and preferences without recourse to evidence and relevant analysis" (p. 18). Having a thoughtful artifact from practice as a focal point can connect the teachers to the discussion and to different ideas. For example, the use of artifacts such as student work affords the opportunity for teachers to question their evaluations about student learning, as well as think about ways to observe students (Akiba, 2007). As teachers craft a response to the artifact, they position themselves publicly, selecting certain variables to notice and not to notice, and this response provides the impetus for beginning a discussion that creates connections in understanding. Issues of difference or inconsistency in teachers' thinking may take place among colleagues, and these too are part of the learning process. Such negotiations around meaning may delay agreement and hence slow down reaching consensus; however, these discussions are necessary to make "issues of purpose, worth, and effectiveness available as matter of collective concern" (Little, 2003, p. 926). This collective meaning making defines the community of practice. For example, Kazemi and Franke (2004) found that the use of "artifacts supported the development of shared language that, in turn, contributed to the construction of workgroup meeting practices" (p. 230).

Life-long learner

The final condition necessary to support teacher learning within Lesson Study is the opportunity to continue to develop and grow in learning. Experience alone is not always a good teacher; you can teach for 20 years but teach the same thing for each of those 20 years (Ball & Cohen, 1999). Unless you have other ideas, models, or mentors to help you re-evaluate your actions and position, experience alone may not give you a different perspective. Multiple perspectives and then reconciliation of those perspectives in the light of evidence can create shared meaning for a community of practice (Kazemi & Franke, 2004). Creating shared meaning comes as the result of articulating new relationships between issues being discussed. Discovering new relationships feeds the desire for more learning and passion to keep finding new ways to approach teaching materials and communicate ideas (Yoshikawa, 2007). From this process of making meaning within a community, inquiry as a stance takes shape. Providing teachers with access and opportunity to express understanding, teachers need to be able to make connections between personal understanding and/or practice and shared meanings developing in the group. As these connections become explicit, teachers then need many opportunities to develop what they understand through further inquiry. Collectively, these conditions develop an individual's capacity to frame and reframe their thinking and develop their inquiry stance within the broader social and political context of the group. To create such conditions for learning, time is needed to reflect on the learning process itself and to cultivate and realign working norms as needed. These norms must continually embrace each individual's contribution while carrying the larger vision of developing both individual and collective teacher learning.

The inquiry stance during Lesson Study is a complex system of activity and communication. It creates conditions that develop authentic research questions, surfaces important rationales, values and integrates colleagues' and others' perspectives, cultivates a keen interest and expertise to understand and articulate teaching and learning, attunes the senses to see and hear student learning and colleagues' learning, builds skills and habits to record and report on classroom interactions, remains open to potential new relationships between issues being discussed, seeds curiosity for further inquiry, and ultimately ignites desire and courage to integrate new ideas in practice. This list is not exhaustive or fully comprehensive; rather it attempts to provide a framework to reflect on what is often invisible.

In the upcoming Lesson Study mathematics cases presented in this book, the inquiry stance will be fleshed out and made visible by the Lesson Study

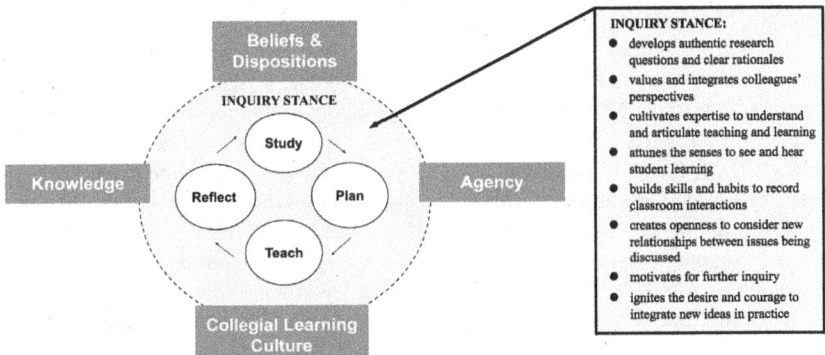

Figure 1.3 Lesson Study's inquiry stance

practitioners. We hear directly from those involved in Lesson Study about how their cycles of inquiry impact their understanding of teaching and learning, specifically through their lesson plans and artifacts we understand what and how they learn together. We see how their focus on multiple perspectives and the reconciliation of those perspectives in the light of evidence creates shared meaning for Lesson Study teams. We see that creating this kind of shared meaning comes as the result of articulating new relationships between the issues being discussed. We also see how discovering new relationships feeds each teacher's desire for more learning and ignites the passion to keep finding new ways to approach teaching. At the end of each case, the teacher-learner is transformed by new ways of knowing, and their desire and courage to try out what they have discovered in the classroom are ignited.

Note

1   The research reported here was supported by the Institute of Education Sciences, U.S. Department of Education, through Grant R305A150043 to Mills College. The opinions expressed are those of the author and do not represent views of the Institute or the U.S. Department of Education.

References

Akiba, K. (2007, March). Japanese teachers' learning system in school: Collaborative knowledge-building through lesson study. Paper presented at the Seoul University, University of Tokyo Joint Conference, Seoul, Korea.

Annenberg Institute for School Reform. (2005). *Professional learning communities: Professional development strategies that improve instruction.* Brown University.

Ball, D. L., & Cohen, D. K. (1999). Developing practice, developing practitioners: Toward a practice-based theory of professional education. In G. Sykes & L. Darling-Hammond (Eds.), *Teaching as the learning profession: Handbook of policy and practice* (pp. 3–32). Jossey-Bass.

Boylan, M., Coldwell, M., Maxwell, B., & Jordan, J. (2018). Rethinking models of professional learning as tools: A conceptual analysis to inform research and practice. *Professional Development in Education, 44*, 120–139.

Bryant, D., & Driscoll, M. (1998). *Exploring classroom assessment in mathematics: A guide for professional development.* National Council of Teachers of Mathematics.

California State Department of Education. (1992). *The mathematics framework.* Author.

Chazan, D., & Ball, D. L. (1999). Beyond being told not to tell. *For the Learning of Mathematics, 19*(2), 2–10.

Chong, A. H., & Kong, C. A. (2012). Teacher collaborative learning and teacher self-efficacy: The case of lesson study. *The Journal of Experimental Education, 80*, 263–283.

Cochran-Smith, M. (2006). Series forward. In C. W. Langrall (Ed.), *Teachers engaged in research: Inquiry into mathematics classrooms in grade 3–5.* National Council of Teachers of Mathematics.

Cochran-Smith, M., & Lytle, S. L. (1993). *Inside/outside: Teacher research and knowledge.* Teachers College, Columbia University.

Darling-Hammond, L. (1998). Teacher learning that support student learning. *Educational Leadership, 55*(5).

Dudley, P. (2013). Teacher learning in Lesson study; What interaction- level discourse analysis revealed about how teachers utilized imagination, tacit knowledge of teaching and fresh evidence of pupils learning, to develop practice knowledge and so enhance their pupils' learning. *Teaching and Teacher Education, 34*, 107–121.

Dufour, T. (2001). How to launch a community. *Journal of Staff Development, National Staff Development Council, 22*(3), 50-51.

Fernandez, C., Cannon, J., & Chokshi, S. (2003). A U.S.-Japan lesson study collaborative reveals critical lenses for examining practice. *Teaching and Teacher Education, 19*(2), 171–185.

Franke, M. L., & Kazemi, E. (2001). Teaching as learning within a community of practice. In T. Wood, B. S. Nelson, & J. Warfield (Eds.), *Beyond classical pedagogy*. Lawrence Erlbaum Associates.

Fuson, K. C., Kalchman, M., & Bransford, J. D. (2005). Mathematics understanding: An introduction. In S. Donovan & D. Bransford (Eds.), *How students learn: Mathematics in the classroom*. National Research Council.

Heaton, R. (2000). *Teaching mathematics to the new standards*. Teachers College Press.

Huang, R., & Shimizu, Y. (2016). Improving teaching, developing teachers and teacher educators, and linking theory and practice through lesson study in mathematics: An international perspective. *ZDM, 48*, 393–409.

Kazemi, E., & Franke, M. (2004). Teacher learning in mathematics: Using student work to promote collective inquiry. *Journal of Mathematics Teacher Education, 7*, 203–235.

Lewis, C. (2002). *Lesson study: A handbook of teacher-led instructional change*. RBS Publishing.

Lewis, C., Friedkin, S., Emerson, K., Henn, L., & Goldsmith, L. (2019). How does lesson study work? Towards a theory of lesson study process and impact. In R. Huang, A. Takahashi & J. P. da Ponte (Eds.), *Theory and practice of lesson study in mathematics* (pp. 13–38). Springer.

Lewis, C., & Perry, R. (2015) A randomized trial of Lesson Study with mathematical resource kits: Analysis of impact on teachers' beliefs and learning community. In: J. Middleton, J. Cai, & S. Hwang (Eds.), *Large-scale studies in mathematics education. Research in mathematics education*. Springer. https://doi.org/10.1007/978-3-319-07716-1_7

Lewis, C., Perry, R., & Hurd, J. (2004). A deeper look at lesson study. *Educational Leadership, 61*(5), 18–23.

Lewis, C., Perry, R., & Murata, A. (2006). How should research contribute to instructional improvement? The case of lesson study. *Educational Researcher, 35*(3), 3–14.

Lewis, C., & Tsuchida, I. (1998). A lesson is like a swiftly flowing river: Research lessons and the improvement of Japanese education. *American Educator, 22*, 14–17 & 50–52.

Lewis, J. M., Fischman, D., Riggs, I., & Wasserman, K. (2013). Teachers learning in lesson study. *The Mathematics Enthusiast, 10*(3), 583–619.

Lieberman, A., & Wood, D. (2002). *Inside the national writing project: Connecting network learning and classroom teaching*. Teachers College Press.

Little, J. (2003). Inside teacher community: Representations of classroom practice. *Teachers College Record, 105*, 913–945.

Lord, B. (1994). Teachers' professional development: Critical colleagueship and the role of professional communities. In N. Cobb (Ed.), *The future of education: Perspectives on national standards in America* (pp. 175–204). College Entrance Examination Board.

Lortie, D. (1975). *School teacher: A sociological study*. University of Chicago Press.

LSRG (2021a). *What is Lesson Study?* Retrieved July 5, 2021, from https://lessonresearch.net/about-lesson-study/what-is-lesson-study/

LSRG (2021b). *Teacher learning during Lesson Study*. Retrieved July 5, 2021, from www. lessonresearch.net/teacher-learning

LSRG (2021c). *Growth dimension of teacher learning in Lesson Study*. Retrieved July 5, 2021, from www.lessonresearch.net/teacher-learning/growthdimensions

National Council of Teachers of Mathematics. (1989). *Curriculum and evaluation standards for schools mathematics*. Reston, VA: National Council of Teachers of Mathematics.

National Council of Teachers of Mathematics. (2000). *Principles and standards for school mathematics*. Reston, VA: National Council of Teachers of Mathematics.

National Council of Teachers of Mathematics. (2001). *Navigations-steering through principles and standards*. Reston, VA: National Council of Teachers of Mathematics.

National Research Council. (2000). *How people learn: Brain, mind, experience, and school* (expanded edition). National Academy Press.

Pellegrino, J. W., Chudowksy, N., & Glaser, R. (2001). *Knowing what students know: The science and design of education assessment*. National Academy Press.

Puchner, L. D., & Taylor, A. R. (2006). Lesson study, collaboration and teacher efficacy: Stories from two school-based math lesson study groups. *Teacher and Teacher Education, 22*, 922–934.

Schipper, T. M. (2019). Teacher professional learning through Lesson Study: an examination of Lesson Study in relation to adaptive teaching competence, teacher self-efficacy, and the school context. *Rijksuniversiteit Groningen*. https://doi.org/10.33612/diss.98636764

Schon, D. (1987). *The reflective practitioner: Toward a new design for teaching and learning in the professions*. Jossey-Bass Inc.

Sibbald, T. (2009). The relationship between lesson study and self efficacy. *School Science and Mathematics, 109*, 450–460.

Silver, E. A., Ghousseini, H., Gosen, D., Charalambous, C., & Font Strawhun, B. T. (2005). Moving from rhetoric to praxis: Issues faced by teachers in having students consider multiple solutions for problems in the mathematics classroom. *Journal of Mathematical Behavior, 24*, 287–301.

Smith, M. S., Stein, M. K., & Silver, E. A. (2005). *Improving instruction in geometry and measurement: Using cases to transform mathematics teaching and learning* (Volume 3). Teachers College Press.

Snow-Gerono, J. (2005). Professional development in a culture of inquiry: PDS teachers identify the benefits of professional learning communities. *Teaching and Teacher Education, 21*, 241–256.

Stein, M. K., & Lane, S. (1996). Instructional tasks and the development of student capacity to think and reason: An analysis of the relationship between teaching and learning in reform mathematics project. *Educational Research and Evaluation, 2*(1), 50–80.

Stigler, J. W., & Hiebert, J. (1999). *The teaching gap: Best ideas from the world's teachers for improving education in the classroom*. Summit Books.

U.S. National Commission on Excellence in Education. (1983). *A nation at risk: The imperative for educational reform*. The National Commission on Excellence in Education.

Wilson, S. M. (2003). *California dreaming: Reforming mathematics education*. Teachers College Press.

Yoshida, M. (1999). Lesson Study: A case study of a Japanese approach to improving instruction through school-based teacher development (Unpublished Doctoral dissertation). University of Chicago.

Yoshikawa, Y. (2007, July 2). *Pre-service education in Japan*. Paper presented at the Lesson Study Immersion Program at the University of Yamanashi, Japan. Translation provided by Global Education Resources.

Learning from Lesson Study as part of the planning team

Thomas McDougal

It is impractical, if not impossible, for pre-service teachers to learn everything they might need to know as teachers of mathematics. For example, although nearly all elementary teachers feel well prepared to teach number and operation, only about half feel well prepared to teach geometry or measurement and data representation (Malzahn, 2019). And, our understanding of the best ways to help students learn mathematics will always be evolving. So, it is important that teachers continue to grow – that they expand both their content knowledge and their pedagogical content knowledge (e.g. Hill, Ball, & Schilling, 2008; Hill, Rowan, & Ball, 2005), and that they refine their teaching skills throughout their career.

Of course, the goal of most teacher professional development is to help teachers grow in ways that will impact their everyday teaching, and thus improve student learning. What makes Lesson Study unique is that it "explodes" key components of teaching – specifically planning and reflecting on lessons – slowing the process down to provide time for teachers to engage more deeply and thoughtfully than normally possible. In fact, as this section will argue, Lesson Study provides an abundance of learning opportunities for the participating teachers during the planning process, at the lesson itself, and at the post-lesson discussion.

An important step in Lesson Study, crucial to teacher learning, is *kyouzai kenkyuu* (Watanabe, Takahashi, & Yoshida, 2008). The literal translation of this phrase is "[the] study of [raw] instructional materials." The idea of "instructional materials" includes manipulatives and other tools (e.g. calculators, maps, videos, scientific equipment), as well as tasks or problems and their contexts. Textbooks are collections of *kyouzai*.

Kyouzai kenkyuu consists of an investigation and consideration of available *kyouzai* with respect to specific instructional goals and the current state of the students. Thus it is important to include, as part of Lesson Study, a careful read of the standards to guide our instructional goal-setting. For example, one team discovered by studying the standards that estimating the area of "spots" (irregular shapes) – which was in their curriculum and with which students struggled – was not actually in the standards. This realization led them to focus their unit and research lesson on helping students understand how they could "use what they

DOI: 10.4324/9781003230915-3

know about multiplication and rectangles to find area of rectangles more effi-ciently." The team also decided to change their units on area:

> We made significant changes to the gr. 3 and 4 area lessons that better align lessons to CCSS and eliminate lessons that required painstaking combining and estimating/counting to estimate the area of irregular shapes. . . . [H]ad we not engaged in Lesson Study, we likely would still be teaching the Area [*sic*] unit the same way.
>
> (Ditto et al., 2019)

To understand the current state of the students, teachers will usually review not only the standards for the target grade level but also for bracketing grades, to know what their students ought to have learned before and what may be left to the future. Then they will study the textbooks, again for both the target grade and ear-lier grades, to know more about what *kyouzai* the students will have been exposed to prior to the unit they are planning. Jana Morse and her team (Chapter 2.2) discovered that their curriculum focused almost exclusively on "result-unknown" problems in join and separate contexts, which helped them understand why their students tended to dive into computation without thinking carefully about what operation they needed to do.

It can also be illuminating to study other textbooks to see how they sequence the learning and what tasks and contexts they provide. Many if not all of the teach-ers contributing to this volume have gained a deeper understanding of content, or new ideas about teaching, by studying translations of Japanese textbooks (e.g., Fuji & Majima, 2020). For example, Lerner and his colleagues discovered that the Japanese text they looked at devoted eight lessons to the subject of division with remainders, in contrast to just one or two in the American texts, and included a lesson on why the remainder must always be less than the divisor, a topic not addressed at all in the American texts (Lerner, 2020). By comparing the textbook series used at their school with another, Whidden, Roman, and Morales (2021) realized that their series did not adequately prepare students to understand tens as a countable unit, which contributed to the difficulties their students had with two-digit addition in grade 2.

Finally, teachers will often consult published journal articles and other research during their study and planning. Lewis et al. (2006), citing Elmore (1996), write:

> Richard Elmore has long argued that U.S. education suffers not so much from an inadequate *supply* of good programs as from a lack of *demand* for good programs on the part of practicing educators. He notes that "the primary problem of scale is understanding the conditions under which people working in schools seek new knowledge and actively use it to change the fundamental processes of schooling."
>
> (p. 280)

As a result of their research, teachers can deepen their understanding of content. This can happen even for content the teachers assume they understand well. For example, in Chapter 2.1, Brigid Brown describes how, prompted by a book they consulted, her team worked through various types of subtraction problems and discovered that some types were more difficult to represent with manipulatives than others, and that they came up with different equations to represent them. Teachers may also discover common misconceptions that they hadn't previously been aware of, leading them to change their teaching. For example, Singer et al. (2019) learned from their research that using language such as "3 over 4" for the fraction ¾ could reinforce a misconception that a fraction represents two values instead of just one. (See also her Chapter 4.3, in this volume.) When this connects directly to a lesson they are going to teach, it can help them appreciate the usefulness of research and be more inclined to seek it out in the future.

Thus we see that *kyouzai kenkyuu* is a way for teachers to build their content knowledge and their pedagogical content knowledge. The process is supercharged in Lesson Study, but in Japan *kyouzai kenkyuu* is a normal part of lesson planning. Minimally, it includes a careful consideration of the tasks and problems in one's textbook, with consideration of the learning goals and current state of the students. So the most important long-term benefit teachers can gain from their *kyouzai kenkyuu* in Lesson Study may be to learn to do it better – or at all – in their daily and weekly planning.

Lesson Study also enables teachers to learn new ideas about what it means to plan a lesson. One of the standard steps in planning a research lesson is to anticipate student solutions. For many teachers, this is a new idea: in their prior work, they were accustomed to focusing on what *they* will do, the examples *they* will show, the tasks *they* will present to students. We have heard many teachers report, as Aubrey E. Perlee does in Chapter 3.3, that their experience with Lesson Study has led them to anticipate student responses in their daily lesson planning in a way that they didn't do before.

The detailed thinking that goes into planning a research lesson, and frequently the comments of outside observers, shine a light on other factors that can significantly impact student learning. How should a task or problem be presented – on a handout, on a poster, or acted out? Should students work in groups, in pairs, or independently? What kinds of manipulatives are most helpful? Should every student have their own manipulatives, should they be shared, or is it sufficient to have an enlarged set that can be used on the board? How should the teacher handle a student who is solving a problem incorrectly? How long should students be allowed to work on a task before it is discussed? What kinds of errors or misconceptions are worth discussing? These are important decisions, but teachers rarely have time to think carefully about them when planning their lessons. Lesson Study provides the time as well as the collaborative environment for teachers to think about and discuss the relative merits of these and many other decisions that have to be made.

One of the details that is addressed as a normal part of Lesson Study is the board plan. In *The Teaching Gap*, Stigler and Hiebert (1999) noted that all the Japanese lessons they recorded as part of the TIMSS video study used the chalkboard, in contrast to many U.S. lessons that used an overhead projector. This was not merely a matter of different preferences in visual aid, but reflected a deeper difference in purpose. Whereas U.S. teachers mostly used the overhead to focus students' attention, Japanese teachers used the board "to provide a record of the problems and solution methods and principles that are discussed during the lesson" (p. 74). The Japanese even have a technical term for board work, *bansho*, and board work is sometimes a topic during the post-lesson discussion of a research lesson.

Creating a board plan is a standard part of Lesson Study – all of the lessons described in this volume had a plan for using the board, and Aubrey E. Perlee discusses the roles of the board in her work in Chapter 3.3. For teachers who haven't previously thought very much about how to use the board, Lesson Study provides an opportunity to learn how to use a board more effectively to support student learning.

During the lesson, while the teacher is busy leading the lesson, other members of the planning team record their observations. Because they helped plan the lesson and therefore know pretty well what the teacher is going to do, they are free – and indeed obligated – to focus on how the lesson is impacting the students. Lesson Study thus provides an opportunity for many teachers to develop new skills for observing students.

For too many teachers, their last opportunity to observe someone else's lesson was during their preservice training, at which time they likely focused on what the teacher was doing in order to learn what they should do when it was their turn. When teaching one's own lesson, one's attention is necessarily divided among all the students. As observers, however, the members of the planning team are each free to focus on one or just a few students, watching what they do, eavesdropping on what they say. By doing so, they can learn more about student misconceptions. For example, a student might write something incorrect, then erase it and copy what is written on the board, or, conversely, never notice a discrepancy between what he or she wrote and what is on the board (for example, Lewis et al., 2013). The teacher might never know about the misconception, but the observer does. An overheard "Why are we doing this?" can help a teacher think about how to design lessons so that the purpose of a task or activity is clear to students. Or, watching a student finish or get stuck early and sit idly for a long time while the teacher helps others can lead one to think more critically about the wisdom of helping individual students (McDougal, 2016). Furthermore, by discovering what can be learned by observing closely, teachers can learn to be better observers within their own classroom. This is why Japanese educators say that Lesson Study gives teachers "eyes to see students" (Lewis, 2002).

After the lesson, misconceptions are brought out in the discussion. Here is where having outside observers, who come with fresh eyes, can be especially helpful. Such observations can lead teachers to significantly change how they teach. Although the discussion normally focuses on the lesson and the students, sometimes comments will help a teacher adjust his or her teaching behaviors, like talking too fast or loudly (Lewis & Tsuchida, 1998).

A good final commentator will also bring important insights to the post-lesson discussion. These might be an analysis of the standards, connections to content in other grades, comments on the research proposal (lesson plan), suggestions for improving future proposals, observations from the lesson itself along with her or his analysis, comments regarding the research theme and how this lesson addressed it, and suggestions for future work (Takahashi, 2014). Many of the sections in this volume identify important insights that the authors gained from the final commentator at the lesson.

Perhaps most important, however, teachers can learn to teach in a new way. All of the teachers who contributed to this volume learned to teach mathematics through problem solving through Lesson Study. In their classes, students provide mathematical reasoning in every lesson, and compare and contrast solution strategies in most of them. In contrast, less than half of U.S. elementary and middle school teachers report having students provide mathematical reasoning in most or all of their classes, and less than a quarter have their students analyze the thinking of others or compare and contrast different solution strategies (Banilower et al., 2018). Stigler and Hiebert (1999) credit Lesson Study in Japan for supporting a nationwide shift to teaching math through problem solving.

How can Lesson Study support teachers in making such a dramatic shift in teaching? As previously noted, Lesson Study provides a structured time for teachers to figure out what a lesson built around problem solving might look like for specific content. In addition, "The process of lesson study appears to support risk-taking in implementing new approaches to teaching and learning by providing a collegial and safe environment" (Lewis et al., 2013). Teachers are more willing to try a new approach because the shared responsibility for the lesson makes them feel less personally vulnerable if it goes badly. But the shared ownership of the lesson, as well as the presence of observers who have read the lesson, creates useful pressure on the teacher of the lesson to carry out the vision of the planning team (see Meghan Smith's account in Chapter 3.2).

Lesson Study focuses on the four most critical tasks of teaching: designing lessons, teaching lessons, observing and analyzing student responses, and reflecting on implications for future lessons. By engaging in Lesson Study, teachers gain content knowledge and pedagogical content knowledge; they learn to design more effective lessons; they gain insights into student thinking; and they are able to try out and practice new pedagogical approaches in a supportive environment.

References

Banilower, E. R., Smith, P. S., Malzahn, K. A., Plumley, C. L., Gordon, E. M., & Hayes, M. L. (2018). *Report of the 2018 NSSME+*. Horizon Research, Inc.

Ditto, C., Lewis, M., McMahon, L., Lynch, J., Nash, M., & Larios, M. (2019). *Using multiple strategies to find area* (Research lesson report). http://LSAlliance.org/Lessons

Elmore, R. F. (1996). Getting to scale with good education practice. *Harvard Educational Review, 66*, 4.

Fujii, T., & Majima, H. (2020). *New mathematics for elementary school* (Trans. A. Takahashi & T. Watanabe). Tokyo Shoseki.

Hill, H. C., Ball, D. L., & Schilling, S. G. (2008). Unpacking 'pedagogical content knowledge': Conceptualizing and measuring teachers' topic-specific knowledge of students. *Journal for Research in Mathematics Education, 39*(4), 372–400.

Hill, H. C., Rowan, B., & Ball, D. L. (2005). Effects of teachers' mathematical knowledge for teaching on student achievement. *American Educational Research Journal, 42*(2), 371–406.

Lerner, J. (2020). Recommendations for teaching division with remainders. *Mathematics Teacher: Learning and Teaching, 13*(8).

Lewis, C. (2002). Does lesson study have a future in the United States? *Nagoya Journal of Education and Human Development,* (1), 1–24.

Lewis, C., Perry, R., Hurd, J., & O'Connell, M. P. (2006). Lesson Study comes of age in North America. *Phi Delta Kappan, 88*(4).

Lewis, C. & Tsuchida, I. (1998). A lesson is like a swiftly flowing river: Research lessons and the improvement of Japanese education. *American Educator, 22*, 14–17 & 50–52.

Lewis, J., Fischman, D., Riggs, I., & Wasserman, K. (2013). Teacher learning in Lesson Study. *The Mathematics Enthusiast, 10*(3).

Malzahn, K. A. (2019). *2018 NSSME+: Status of elementary school mathematics.* Horizon Research, Inc.

McDougal, T. (2016). It's not the teacher's job to teach the students! *Blog Post.* www.lsalliance.org/2016/01/not-teachers-job-teach-students/

Singer, L., Velasco, M., Kim, D., Cadena, M. (2019). Adding fractions. *Research Lesson Proposal.* http://LSAlliance.org/Lessons

Stigler, J. W., & Hiebert, J. (1999). *The teaching gap: Best ideas from the world's teachers for improving education in the classroom.* The Free Press.

Takahashi, A. (2014). The role of the knowledgeable other in Lesson Study: Examining the final comments of experienced Lesson Study practitioners. In *Japanese Lesson Study: A model for whole-school teacher professional learning* (1st ed., Vol. 16). Mathematics Teacher Education and Development, Mathematics Education Research Group of Australasia, Inc.

Watanabe, T., Takahashi, A., & Yoshida, M. (2008). *Kyozaikenkyuu:* A critical step for conducting effective Lesson Study and beyond. In *AMTE monograph 5, inquiry into mathematics teacher education* (pp. 131–142).

Whidden, H., Roman, I., & Morales, A. (2021). 2-digit addition with regrouping. *Research Lesson Proposal.* http://LSAlliance.org/Lessons

What can we learn from observing a research lesson?

Tad Watanabe

Teachers and other educators participate in a Lesson Study cycle in different capacities. As Takahashi and Yoshida (2004) described, the core participants are those who are in the planning group that plans the research lesson. However, there are others who are outside of this core group who participate in Lesson Study. For those who are outside of the planning team, the most visible parts of a Lesson Study cycle are the publicly taught research lesson and post-lesson discussion, that is, the Teach and Reflect steps in the model shared by Friedkin earlier (see Figure 1.1 on p. 5). A planning team is usually small, with maybe up to six to eight teachers, and typically only one teacher in the group gets to teach the observed research lesson. However, observing a research lesson and engaging in the post-lesson discussion afford many more teachers, even those outside of the planning team, a valuable professional learning opportunity. After the idea of Lesson Study was widely introduced to the US mathematics education community by Stigler and Hiebert (1999), some teams in the United States tried to implement their Lesson Study cycles without involving teachers outside of the planning team for a variety of practical reasons. However, Lewis and Tsuchida (1998) discuss how participation in this observation and discussion of research lessons influenced Japanese teachers' professional learning. Thus, observing a research lesson and participating in the post-lesson discussion are significant opportunities for teacher learning.

Before the introduction of Lesson Study, professional learning opportunities for teachers in the United States rarely involved observation of actual lessons and discussion afterward – and I suspect this was true in many other countries as well. A typical professional learning opportunity might involve a workshop or seminar in which teachers learn new ideas, and sometimes these sessions might include classroom videos to illustrate the ideas. The participating teachers are then expected to implement the ideas in their own classrooms. Occasionally, a workshop may include opportunities for participating teachers to engage in micro-teaching in which one teacher acts as the teacher of the lesson and others play the role of students. But in my experience, there seems to be much more emphasis on "doing" and "listening" than observing and reflecting in most typical professional learning activities. In fact, there seems to be an implicit belief that there is only so much we can learn from observing. Is it any different in Lesson Study?

DOI: 10.4324/9781003230915-4

The original Japanese word that has been translated as "study" in Lesson Study can also be translated as "research." In fact, the public lessons in Lesson Study cycles are translated as "research" lessons even though the same Japanese word is used for both. Thus, the original Japanese phrase could have been translated as "lesson research." Therefore, as Friedkin earlier discussed, Japanese teachers working with American teachers who are trying to implement Lesson Study at their school encouraged the American peers "to see themselves as researchers conducting empirical examination, organized around asking questions about practice and designing classroom experiments to explore these questions" (Fernandez, Cannon, & Chokshi, 2003, p. 173), because for Japanese teachers, they are indeed engaged in research. In a research lesson, the planning team presents their hypothesis on how to tackle a challenge in mathematics teaching and learning they have identified. Thus, a research lesson is like a research report presentation at a professional meeting. Therefore, just as scholars learn from participating in other scholars' research report presentations, teachers can learn from observing and discussing a research lesson developed by other teachers in a Lesson Study (or "Lesson Research") cycle.

So, what can teachers learn from observing and discussing a research lesson? Imagine a group of middle grade teachers observing a Grade 6 (12-year-old) research lesson on finding the area of parallelogram. The research theme for this planning group was "designing instruction that will lead to robust understanding of the area formulae for quadrilaterals and triangles." The research lesson was the second lesson in the unit on the area of triangles and quadrilaterals. In the first lesson, the class explored how they might make use of what they had already learned, that is, how to calculate the area of rectangles and squares, to find the area of parallelogram shown in Figure 1.4. Many teachers are familiar with a lesson like this, and as expected, many students were able to find the area by transforming the given parallelogram into a rectangle, usually by cutting off a triangle on one end of the parallelogram and re-arranging it to make a rectangle. The resulting rectangle had the horizontal dimension that was equal to the horizontal side of the parallelogram, and the vertical dimension that was equal to the length of the perpendicular segment used to cut off the triangle. They labeled the horizontal side of the parallelogram as the base, and the segment used to cut off the triangle as the height. Thus, the class concluded that the area of this parallelogram could be calculated by the formula *base × height*.

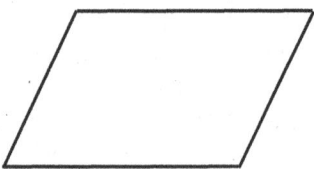

Figure 1.4 The parallelogram used in the first lesson of the unit

In the research lesson, the teacher asked students to find the area of the parallelogram shown in Figure 1.5. As students tried to apply what they learned previously, they ran into a problem. A student was heard saying, "the height stops here and doesn't reach the base." Some tried to use the slanted side as the base, but they realized that they can't determine the length of the base – nor the height. The teacher then asked students to think about ways to figure out the area of this parallelogram. They were provided with the copies of the parallelogram, printed on different colored papers which they were free to manipulate/cut.

After about seven minutes of independent problem-solving time, the teacher started asking some students to post their solution approaches on the board, where large copies of the parallelogram were posted. In some cases, the teacher asked students to only show what they did on the drawing of the parallelogram, and with some others he asked them to write only the equations showing the calculations they did. Altogether, six different ideas (see Figure 1.6) were posted on the board when the class started the whole class discussion. The class first made sure that they understood what each student did. For some that had only equations, the teacher asked if others could figure out what the students did so that the shown equations could be used to calculate the area of the parallelogram. For others that only show what they did on the picture of the parallelogram, the teacher asked the class what equations would show the calculations necessary to find the area.

After the class made sense of each idea, the teacher asked the students to look for similarities and differences among those ideas. Students noticed that from four of the ideas (Figure 1.6 a, b, c, and e) they can show that the area of the parallelogram is the same as the area of the rectangle on the base that reached to the parallel line that contained the opposite side of the base (see Figure 1.7).

The class concluded that if they interpreted the height of the parallelogram as the distance (or width) between the parallel lines containing the base and the opposite side, we can still use the formula *base* × *height* to find the area. In other

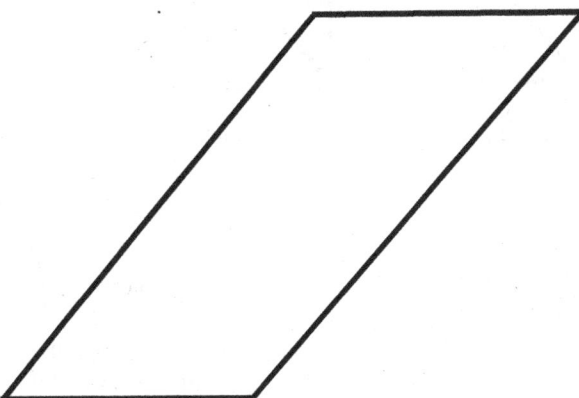

Figure 1.5 The parallelogram used in the research lesson

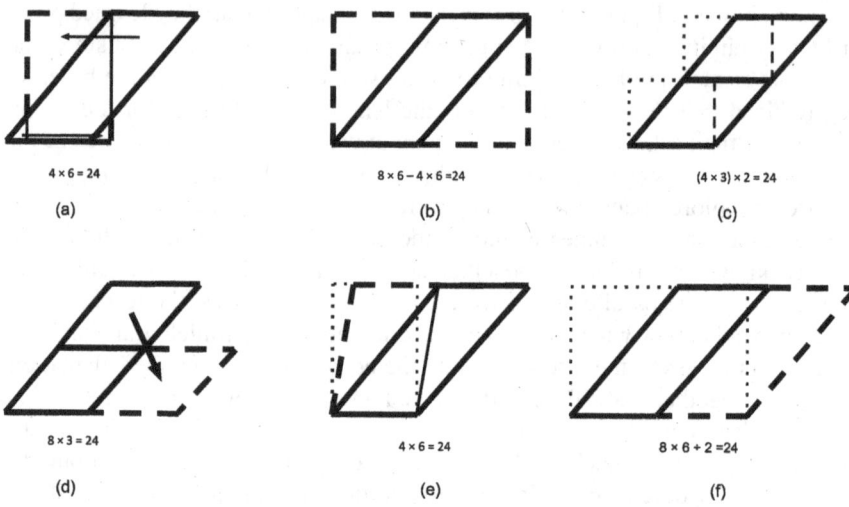

4 × 6 = 24

(a)

8 × 6 − 4 × 6 =24

(b)

(4 × 3) × 2 = 24

(c)

8 × 3 = 24

(d)

4 × 6 = 24

(e)

8 × 6 ÷ 2 =24

(f)

Figure 1.6 Six solutions presented by the students

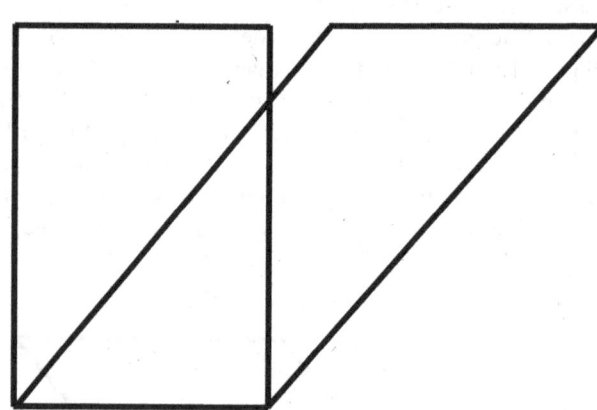

Figure 1.7 The parallelogram and the related rectangle

words, the segment representing the height may be drawn completely inside the given parallelogram, intersecting a side, or even completely outside of the parallelogram (see Figure 1.8).

During the post-lesson discussion, some teachers expressed their amazement on how comfortably the students cut out the copies of the parallelogram and moved them around to help them think about ways to figure out the area of the parallelogram. It was noted that the students have done many geometry explorations where

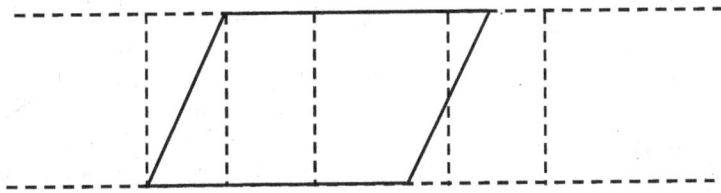

Figure 1.8 Segments representing *height* may be drawn in different places

they cut the given shapes and re-arrange them to create new shapes starting in Grade 3. The planning team also shared that the students have done an exploration to find the area of an L-shape where they used many of the same strategies – cutting and re-arranging (like in Figure 1.6 a and d), adding on extra pieces to make a rectangle (like Figure 1.6 b), and even doubling (like in Figure 1.6 f). It was also discussed that in the future lessons, students would be asked to draw the segment representing height when the slanted side was considered as the base, since in the first two lessons, the base was always the horizontal side.

After observing this research lesson and participating in the post-lesson discussion, what could those middle-grade teachers have learned from it? Some might have realized that concluding the area formula for parallelogram is *base × height* only after the first lesson, as they have done before, is indeed a premature conclusion. Others might have learned the importance of engaging students in activities in which students must represent their ideas in equations as well as interpreting equations of others. Such discussion in the research lesson might have contributed students to develop a deeper understanding of what *height* of a parallelogram is. Some teachers who may be familiar with five practices (Smith & Stein, 2011) might realize that discussion of student ideas does not always have to be sequential, and asking students to look for similarities and differences across various ideas might be another powerful way to make connections. These are just a few examples of possible learning teachers might have as a result of observing a research lesson and participating in the post-lesson discussion.

When teachers observe other teachers teaching, they can learn many ideas. But, their learning may be haphazard and unorganized. Lesson Study provides a structure that might make teacher learning through observation much more structured and intentional. That is because a research lesson is carefully prepared to address specific teaching and learning challenges that have been identified by the planning teams. Thus, although teachers may still learn something just because they are observing other teachers' lessons (for example, printing the parallelogram on different colored papers as done in the research lesson earlier), research lessons and post-lesson discussion provide opportunities for observing teachers to learn ideas that are related to the research questions of the planning team. Observing what teachers learn from a research lesson and post-lesson discussion covers much of what is known as *mathematics knowledge for teaching* (see Figure 1.9).

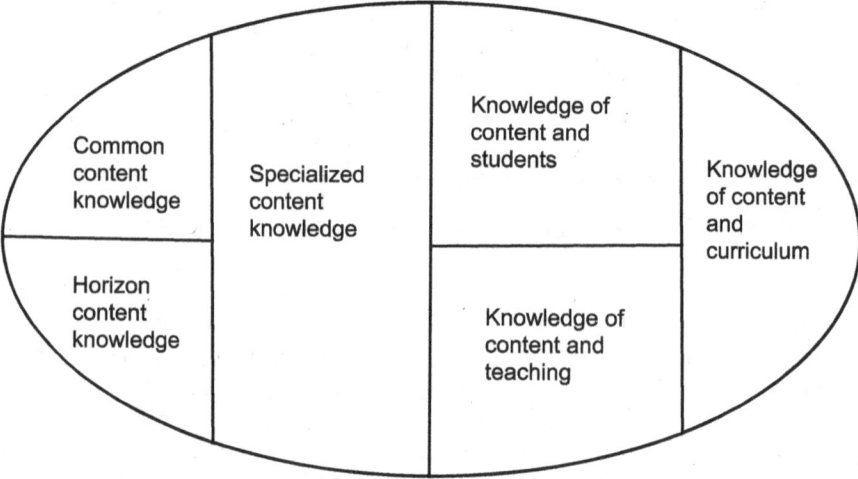

Figure 1.9 Components of mathematical knowledge for teaching

Source: Adapted from Ball, Thames, & Phelps, 2008

What is critical to support teacher learning from observing a research lesson is a carefully developed and thorough research lesson plan, also known as lesson proposal. By reading the plan, which includes the summary of the planning team's *kyouzai kenkyuu* and rationale for the selection of the task and how the team planned to structure the whole class discussion period, observing teachers can develop their own foci for observation. Having prior knowledge of various teaching moves, observing teachers can more accurately assess the effectiveness of those moves, or raise questions based on their observations during the post-lesson discussion.

Although the observers can learn much from observing and discussing a research lesson afterward, observers also contribute to the planning team members' learning as well. A research manuscript is published after it is peer-reviewed. Thus, we can also consider the analogy of a research lesson as the manuscript submitted for review, and the post-lesson discussion as the dialogue between peer reviewers and the research team. I suspect all scholars agree that the final version of their manuscripts is better after they go through this peer review process. In the same way, observers can sharpen and deepen the planning team members' learning as well.

In the following chapters, teachers and educators share what they learned from participating in Lesson Study. Some of the authors were members of the planning team members, while others participated as observers. Their cases will further illustrate not only what teachers may learn from observing and discussing a research lesson but also how observers' and the planning team members' learning is mutually supported.

References

Ball, D. L., Thames, M. H., & Phelps, G. (2008). Content knowledge for teaching: What makes it special? *Journal of Teacher Education, 59*(5), 389–407.

Fernandez, C., Cannon, J., & Chokshi, S. (2003). A U.S.-Japan lesson study collaborative reveals critical lenses for examining practice. *Teaching and Teacher Education, 19*(2), 171–185.

Lewis, C., & Tsuchida, I. (1998). A lesson is like a swiftly flowing river: How research lessons improve Japanese education. *American Educator, 14–17*, 50–52.

Smith, M. S., & Stein, M. K. (2011). *5 practices for orchestrating productive mathematics discussion*. National Council of Teachers of Mathematics.

Stigler, J., & Hiebert, J. (1999). *The teaching gap: Best ideas from the world's teachers for improving education in the classroom*. Free Press.

Takahashi, A., & Yoshida, M. (2004). Ideas for establishing lesson-study communities. *Teaching Children Mathematics, 10*, 436–443.

II What we learned about the contents we teach

DOI: 10.4324/9781003230915-5

"Doing the math"

Word problems in the primary grades – solve "take-from with change unknown" story problems with grade 1 (6- and 7-year-old) students

Brigid Brown

One might think it would be hard to find a topic of common interest for a Lesson Study group spanning three grade levels, two languages of instruction, brand new teachers and veterans, and both a general- and a special-education-inclusion context. That was not the case for this team of primary grade teachers at Acorn Woodland Elementary School, for which I was a facilitator. With resounding conviction, they agreed: our students struggle with subtraction word problems. What causes this struggle? And what learning experiences and tools can we provide to help them make sense of this work? In our research, we would come to discover that a powerful way to answer these questions was to explore subtraction word problems from our students' perspectives, attempting to solve as we imagined they might.

Across these different contexts and among students aged 5 to 8 years old, the teachers had noticed that when our students encounter a word problem, they often rush into calculations without taking time to comprehend the situation in the problem. Often this meant that students would scan the problem, take the two numerals they saw, add them, and name the total as the answer. The team wondered how they could slow students down enough for them to make sense of a problem first and then choose an operation that matched the context of the problem. To do so, students would need to attend to the meanings of the numbers in the problem, especially to the unknown. More broadly speaking, team members wanted to better understand their students' mathematical thinking and to teach their classes in a way that would allow students to experience agency in their learning.

To ground our research in our students' own experience of solving word problems, I led the team in a hands-on activity for exploring word problem types and structures, an activity that the team called "doing the math." Using *Elementary and Middle School Mathematics: Teaching Developmentally* (Van de Walle, Karp, & Bay-Williams, 2016) as our guide, we first read about the different types of addition and subtraction problem types and structures. Then we went through each one and attempted to solve the way we thought our students might, using counters, diagrams, and equations. For reference, a chart similar to the one our

DOI: 10.4324/9781003230915-6

team used is available in the *Mathematics Glossary, Table 1*, of the Common Core State Standards, available online at www.corestandards.org/Math/Content/mathematics-glossary/Table-1/.

Very quickly in the process of "doing the math" ourselves, the team noticed differences in the degrees of difficulty in the problem types and structures. Problems in which the result was unknown were easy to model with counters, and the equations looked familiar. But in problems where the unknown quantity represented the change or the start, we found it was difficult to use the counters to show our work. We were surprised to see that we came up with different equations to represent certain problems as well. For example, in a "take-from" subtraction situation where the change was the unknown, some team members were certain that an equation showing the total minus the known result was the proper representation $(T - R = ?)$, whereas others were convinced that the known result should fall to the right of the equal sign, with the total minus the unknown change to the left $(T - ? = R)$.

Team members reflected on how they'd never have guessed that discussing solution methods for addition and subtraction word problems would have led to such rigorous debate (and frankly, confusion) among a group of adults – these are kindergarten and first grade standards after all! The exercise left team members with a deeper appreciation for the complexities of these problems, and how significantly that complexity can vary among problem types and structures.

The team began to think of word problems as complex texts. Drawing on their literacy work, the team decided to try explicitly teaching students to use visualization with their word problems, just as they did in their reading. As a bridge to this strategy, the team would dedicate a series of lessons to unpacking one word problem at a time, treating the word problem like a storybook. The teachers designed four illustrations to align with the three quantities in their problem plus the question at the end, hypothesizing that this might serve as a productive support for relating the context to the corresponding equation.

Finally, the team considered what discussion questions could guide students to think carefully about the story context and the missing information: "What do we already know? What are we trying to figure out?" These questions could be applied in many lessons throughout the year in order to reinforce the concept of the "unknown" (or "mystery number") in a word problem as a more precise way to think of the "answer."

This process of study left a strong impression on our team of how mathematics teaching in the primary grades can be deceptive in its apparent simplicity. Working through specific word problems as we thought our students might allowed us to experience the "sense-making" required of our students, which was often in contrast to the "answer-getting" we remembered being emphasized in our own early schooling. We saw that even for what seems like a simple problem, "doing the math" could take us a long way towards teaching our students more effectively.

Lesson plan

Title of the lesson: Solve Take-From with Change Unknown Story Problems
Students' school: Acorn Woodland Elementary School, Oakland, CA
Student ages: Grade 1
Instructor: Malia Vitousek
Co-authors: Kristen Brett, Brigid Brown, Shelley Friedkin, Jayme Kritzler, Francisco Llaguno, Maira Lopez, Natalia Ruiz, Tala Sullivan, Malia Vitousek
Date: October 24, 2018
Goal of the lesson: Students will be able to use their chosen strategy effectively to solve the problem on their own. Students will be able to articulate what each number represents.

Learning standards

Use addition and subtraction within 20 to solve word problems involving situations of adding to, taking from, putting together, taking apart, and comparing, with unknowns in all positions, e.g., by using objects, drawings, and equations with a symbol for the unknown number to represent the problem (CCSS Math 1.OA.A.1) (Common Core State Standards Initiative, 2010).

Understand subtraction as an unknown-addend problem. For example, subtract $10 - 8$ by finding the number that makes 10 when added to 8 (CCSS Math 1.OA.B.4) (Common Core State Standards Initiative, 2010).

Flow of the lesson

Introduction

TEACHER: "We have been working on math story problems and today I have a real math storybook to read with you!"

Posing the problem

Teacher reads aloud the story problem: "There are 6 bears swimming. Then some bears went home. Now 2 bears are swimming. How many bears went home?"

Teacher then rereads the story, posting pictures on the board as she goes.

Anticipated responses

Note: Students may use any combination of the following strategies to reach their answer.

- manipulatives
- drawings
- picture bond
- number bond
- equations

R1: Students combine 6 and 2 and name 8 as their answer.

R2: Students use manipulatives or diagrams to separate 6 into 4 and 2, but choose 6 as the answer or are unsure of which number is the answer.

R3: Students use manipulatives or diagrams to separate 6 into 4 and 2 and correctly identify 4 as the answer.

R4: Students use either an addition or subtraction equation but choose 6 or 2 as their answer.

R5: Students use an addition equation to identify 4 as the missing addend. $2 + \_ = 6$

R6: Students use subtraction to identify 4 as the difference. $6 - 2 = \_$

R7: Students use subtraction to identify 4 as the missing subtrahend. $6 - \_ = 2$

Comparing and discussing

Teacher presents two to three strategies that correctly depict the story, so the emphasis in discussion is on understanding how students' strategies connect to the story, not on choosing the correct answer. Ideally one strategy uses manipulatives and one involves an equation, so as to have both concrete and abstract strategies represented in the discussion.

Teacher will ask questions such as:

- What does the (*name specific part of diagram or equation*) represent? Where's this in the story?
- Six whats? (Six bears swimming at first)
- Which numbers did we know? What was the "mystery number"?
- How are these two strategies similar? Different? Where's the (*name specific part of diagram or equation*) in this strategy . . . and in this one?"

Teacher may push students' thinking by

- adding labels according to students' responses
- prompting students to add equations to match their models
- "playing the fool," asking "Ok, so 2 went home?" in order to push students to explain, no, we don't know how many went home.

Teacher may offer the equation $6 - \_ = 2$ under the story book as a way to represent the problem.

Summing up

Planned summary: A picture bond [or number bond] can help us find the mystery number.

Extension problem: There are 9 birds in a tree. Some birds flew away. Now there are 3 birds in the tree. How many birds flew away?

Board plan

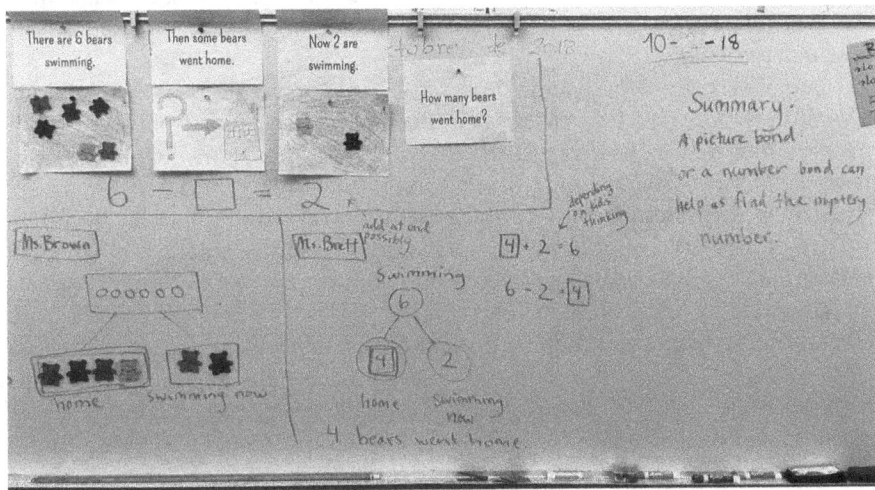

Figure 2.1 Board plan for the lesson

Observations from the lesson

During the lesson, I immediately noticed the routines and norms Malia had been building in her classroom. Malia invited the students into the work of the lesson, asking them "are you up for the challenge?" to which they chorally responded in the affirmative. Students eagerly used sentence frames to practice the language of student-to-student discussion. The social emotional tone of the class was positive, and students seemed excited to solve the problem independently. I noted one student who began his independent work declaring, "let's draw the lake!" creating a clear connection between his work and the context of the story problem.

While most students seemed familiar with the number bond strategy, some seemed to be misunderstanding the tool. For example, some students who found the correct answer of four bears drew a number bond with four at the top, suggesting that they saw the top number as the "answer," not the whole above the two parts. Some students correctly represented the problem with six at the top and four and two below, but then listed their answer as "There are six bears." It seemed like these students might need more explicit instruction in the meaning of the number bond before they would be able to effectively use it to solve or represent problems.

In the post-lesson discussion, observers noted hearing students refer to the pictures in their discussions, which seemed like good evidence that they were making use of the comprehension support the pictures provided. However, one observer noticed that there were a number of students who successfully solved the main problem, but not the extension problem, which was the same problem type and structure without illustrations. She suggested that perhaps having students draw their own illustrations to a word problem might be a productive bridge towards more independence.

Other observers also shared ideas for how to use storybook illustrations in future lessons. One suggested that students might benefit from working in reverse order: if we showed just the pictures, would students be able to construct their own understanding of what might be happening in the problem? Others suggested comparing illustrations for two-word problems with different structures. For example, with two examples side by side, could students explain the differences and similarities both in the story and in the corresponding equations?

In her final commentary, Robin Lovell, a district mathematics coach, addressed the team's goal of making lessons more student-driven by suggesting ways we might facilitate high-level discussions and encourage student-to-student talk. For example, we could compare two students' solutions side by side, and ask: "These two students both solved the problem successfully, but one used addition and one used subtraction. How is that possible? How can two different operations help us arrive at the same answer?" Another discussion might center on the work of two students who wrote the same equation but chose different answers: "How is it that the same equation can be interpreted to have two (or more!) different answers? And how do we know which one is correct?"

It's striking how, using just one- or two-word problems, such a variety of rich, layered, and thought-provoking experiences are possible, when employing student discussion and high-level questioning. This input confirmed the team's thinking that by examining a single problem closely they could help their students attend to meaning and make sense of challenging word problems.

What I learned

Solving different addition and subtraction word problems and studying word problem types and structures gave our team greater insight into the nature of the specific challenges each problem posed. As one team member explained, "Even though it's just first grade math, there's a lot going on cognitively for students that we can take for granted if we don't do the math ourselves."

During our "do the math" activity, the group hit on a crucial moment of uncertainty when we explored a certain problem type: take-from subtraction with change unknown. The problem read something like this:

Brigid had 12 stickers. She gave some to Francisco. Now she has 8 stickers. How many did she give to Francisco?

Everyone attempted the task individually, modeling with counters and writing an equation to match the problem. When we shared our work, two different equations surfaced: $12 - 8 = \_$ and $12 - \_ = 8$. The ensuing debate over which equation was "right" became surprisingly impassioned, and it seemed we might not be able to reach consensus.

Referring back to our text, we found insight into our confusion in the description of semantic and computational equations. According to Van de Walle et al., a semantic equation is one in which "the numbers are written in the order that follows the meaning of the story problem" (p. 157). In contrast, a computational equation is an equivalent equation "which isolates the unknown quantity and would be used if you were to solve this equation with a calculator" (p. 157) In this case, $12 - \_ = 8$ was the semantic equation and $12 - 8 = \_$ was the computational equation.

One team member reflected on the impact this experience had on her. She said that when we first looked at the word problem in the research lesson, it was really hard for her to imagine that anyone might solve it a different way than she had. She went on to say that had she not gone through the Lesson Study process, this lesson would probably have looked like "solve it using 6 minus 2 because that's how you get the correct answer." Now she saw how that could limit students' opportunity to build understanding. "Doing the math" led this team member to reflect on the value of organizing a lesson around exploring and comparing various solution methods to a single problem in order to build students' conceptual understanding.

The Common Core State Standards demand conceptual understanding and problem solving in addition to procedural fluency, whereas many of us teaching now experienced elementary school math classes with a narrower focus on computation and "answer-getting." As teachers, we need a chance to do our own sense-making so that we can better understand the challenges our students face and make instructional choices that help them meet those challenges.

Impacts on my own teaching

"The teacher needs to 'do the math' in order to make the classroom a true math community." This insight, which was shared by a team member in our final reflection meeting, has stayed with me in my work as a teacher leader as I strive to create and nurture math classrooms where students drive the learning. When we engage deeply with the problems our students must tackle and strive to understand the different ways our students make sense, the teacher's interpretation is no longer the center, but instead is one part of a learning community. This preparation helps us to attune our eyes and ears to the sort of sense-making we might expect from our students, allowing us to step into the role of observer and facilitator. And in that way we are better prepared to center our students' ideas, amplifying the sense-making of individual students for the whole class to utilize as they co-construct their understanding of a concept.

This lesson crystalized this understanding for me, especially in terms of the aspect of Teaching through Problem-Solving that I find most difficult: anticipating a student-generated summary. A powerful summary, gleaned from student input at the end of a lesson and guided by the teacher's plan, has the potential to imprint the new learning metacognitively for students, making it readily accessible for future work. It is critically important and, for me, quite challenging. In this work, Lesson Study has been so helpful for me. Among my peers and mentors, I can ask, "Now we've built our own 'grown-up' understanding what makes this problem tricky . . . how can we expect our students to articulate it? What would sense-making conversations sound like among my students?"

And in the case of this lesson, I think the best possible summary was not one of the dozen or so that our team anticipated, but rather it was one articulated by a student. Towards the end of the lesson, a student (we'll call her Nia) stated: "We're trying to find the number that's subtracted." This was an "ah-ha" moment for me. Nia's take on the essential challenge in the lesson is both elegant and precise, in her use of her first grade mathematical language to describe this new learning. Had it not been for our teams' close study of word problem types and structures, it would have been easy to miss the significance of this gem of a statement.

Before "doing the math" ourselves, our team knew this problem was tricky because it was a subtraction word problem. But we did not yet understand the nature of the difficulty: that the unknown falls in the middle of the equation; that the missing number is the subtrahend, or the change, not the result as students are accustomed to seeing. Or, as Nia elegantly explained, "We're trying to find the number that's subtracted." But when we enter a lesson with a deep understanding of the mathematical terrain our students must navigate, we can harness ideas like Nia's to guide the class to co-create their own deep understanding of their math work, building the class's math community around students' own sense-making.

References

Common Core State Standards Initiative. (2010). *Common Core State Standards for Mathematics*. http://www.corestandards.org/Math/

Van de Walle, J. A., Karp, K. S., & Bay-Williams, J. M. (2016). *Elementary and middle school mathematics: Teaching developmentally* (9th ed.). Pearson.

To add or subtract, that is the question

A compare-type problem with the smaller quantity unknown, grade 2 (7- and 8-year-old) students

Jana Morse

Acknowledging the unknown can be a powerful place to start

A feature of Lesson Study that I find most satisfying occurs when a new group assembles to decide on the unit of study. This important first step in the Lesson Study process grants us permission to ask questions without knowing the answers, to make observations without understanding the why behind them, and it encourages us to tackle the aspects of teaching and learning that challenge us most.

We began our research by studying the four addition and subtraction problem types (described in Table 2.1), after noting that second grade students struggled to solve addition and subtraction problems that did not follow the same structure as join-result unknown (the act of adding to a quantity to find an unknown) or separate-result unknown (the act of taking away from a quantity to find an unknown result).

These problem types are helpful when students are first introduced to addition and subtraction in Kindergarten because they allow for students to make sense of the problem by acting out the situation using concrete objects. For example:

Simon had 3 blocks and his friend Jason gave him 2 more. How many blocks does Simon have now?

A student can readily make sense of this problem using concrete objects to represent both Simon's quantity (3) and Jason's quantity (2), before combining both quantities to determine the result.

As our research continued, it became clear why our students held this narrow understanding of addition and subtraction upon entering grade 2; the curriculum focused primarily on finding the "result" in a "join" or "separate" context, so much so that students had come to equate finding the "result" with finding the "answer", despite the fact that the answer depends on the position of the unknown in the problem.

Additionally, we observed that students routinely got stuck when tasked with a problem type that featured relationships instead of actions. In Join and Separate situations, the students are able to model the inherent *actions* using concrete objects or by writing discrete symbols on paper. But Part-Part-Whole– and Compare-type problems feature *relationships* that cannot be modeled in the same

DOI: 10.4324/9781003230915-7

Table 2.1 The four addition and subtraction problem types

Problem type			
Join	*(Result Unknown)* Connie had 5 marbles. Juan gave her 8 more marbles. How many marbles does Connie have altogether?	*(Change Unknown)* Connie has 5 marbles. How many more marbles does she need to have 13 marbles altogether?	*(Start Unknown)* Connie had some marbles. Juan gave her 5 more marbles. Now she has 13 marbles. How many marbles did Connie have to start with?
Separate	*(Result Unknown)* Connie had 13 marbles. She gave 5 to Juan. How many marbles does Connie have left?	*(Change Unknown)* Connie had 13 marbles. She gave some to Juan. Now she has 5 marbles left. How many marbles did Connie give to Juan?	*(Start Unknown)* Connie had some marbles. She gave 5 to Juan. Now she has 8 marbles left. How many marbles did Connie have to start with?
Part- Part- Whole	*(Whole Unknown)* Connie has 5 red marbles and 8 blue marbles. How many marbles does she have?		*(Part Unknown)* Connie has 13 marbles. 5 are red and the rest are blue. How many blue marbles does Connie have?
Compare	*(Difference Unknown)* Connie has 13 marbles. Juan has 5 marbles. How many more marbles does Connie have than Juan?	*(Compare Quantity Unknown)* Juan has 5 marbles. Connie has 8 more than Juan. How many marbles does Connie have?	*(Referent Unknown)* Connie has 13 marbles. She has 5 more marbles than Juan. How many marbles does Juan have?

Source: Adapted from *Children's Mathematics: Cognitively Guided Instruction* (Carpenter, 1999)

way. This new understanding created a need to identify other concrete reasoning models for students to use.

During our deep dive into the research, we were surprised by how much there was to learn about the addition and subtraction problem types. One reason for this could be traced back to our own math education which prioritized calculation and speed over sense making and problem solving (we shared the same memory of the dreaded "word problem" at the end of each chapter). To be considered "good at math", one had to be able to calculate fast and accurately, thus our math time was spent on "drill and kill". Problem solving was but a mere afterthought.

Though the research at times proved humbling, we were excited to drive our own learning and felt rewarded by the process. How could we expand our students' problem-solving ability to include other types of problems with different structures? What tools would our students need to become adept problem solvers, and how could we support them in this endeavor?

These questions formed the basis of our research and drove us in our common goal to design a unit (which included the research lesson) that would support students as they grappled with the most challenging problem type: a comparison problem using "more" when the smaller quantity being compared is unknown. Consider, for example, the following problem:

> Lilly had 15 gumdrops. She had 7 more than Jackson. How many gumdrops did Jackson have?

In this problem, the use of the word "more" can mislead students to choose the incorrect operation.

The aim of our research lesson

We chose to teach a "Compare" problem type that included the word "more" because we wanted to know how to support our students when faced with a problem in which the operation was not apparent. Could we help students leverage their prior knowledge by connecting it to a new problem situation? And how might we incorporate the use of new concrete models to aid students in their reasoning?

Our Lesson Study process

To describe the impact from this Lesson Study process, I think it would be helpful to highlight the backwards design approach we took:

Part I

1   We identified student learning issues.
2   We honed in on our research lesson topic.

Part II

3   We researched the relevant content.
4   We studied the concept progression related to our topic that would inform us not only in the planning of our research lesson but also in the planning of the unit design within which our research lesson would fall.
5   We learned how and why language can mislead students.
6   We identified concrete modeling tools that would be helpful to aid students in their sense-making and reasoning abilities.

Part III

7   We developed a unit (see Table 2.2).

Part IV

8   We created a research lesson.

Learning the concept progression (Common Core State Standards Initiative, 2010) was key in helping to illuminate the complexities embedded *within* and *between* the problem types, while understanding the importance of concrete models was helpful for supporting students as they solved more complex problems.

The value of concept progression

The value of studying standards and content cannot be emphasized enough. What emerges from such a study is an understanding of a progression that reflects the intentional sequencing of concepts built upon previously learned concepts to provide students the support necessary to learn more challenging and sophisticated concepts. Knowing how the concepts progressed was key to us being able to target and address our student needs. Not only did we learn the different levels of complexity *within* each of the four problem types (determined by the place of the unknown) as well as the levels of complexity *between* the problem types (join and separate situations involving action are easier to model than the Part-Part-Whole and Compare situations that involve relationships), we also realized how the language used in problems can mislead students. In a Compare problem when the quantity being compared is unknown (also known as the "smaller" unknown according to Common Core State Standards Initiative, 2010) a student's ability to solve can be undermined if he believes that "more" always means to add. As we wrote in our research lesson document:

> One further consideration is as students reason to understand the different comparison situations, the vocabulary has the potential to be a challenge. For example, students typically associate the word "more" with addition and the word "fewer" with subtraction. The main area of challenge we anticipate for our research lesson is that the word "more" does not always equate to addition.
>
> (Bonner, T. et al., 2016)

The value of discovering new ways to support students as problem solvers

Once we discussed the implications of the "unknown" in any position within each problem type and recognized the importance of introducing concrete models (number lines, bar models, tape diagrams, equations etc.) as reasoning tools for students, we spent time considering how to thoughtfully re-introduce the tape diagram (they had exposure to the tape diagram, the bar model, and the number line in Grade 1) in familiar situations and also in unfamiliar situations in advance of our research lesson. As we further noted in our research document

> In terms of models, we also re-introduced the tape diagram model/ bar model. Students quickly took to this model to make sense of part-part-whole problems; however, many still struggled to construct a correct tape diagram in

subtraction problems or in addition problems involving the start unknown. Numerous days were spent with students wrestling to correctly label their equations and tape diagram models.

(Bonner, T. et al., 2016)

We also found it helpful to encourage students to label the numbers in their diagrams to support their sense-making efforts. Part of sense-making is the ability to contextualize and decontextualize numbers as needed (Common Core State Standards Initiative, 2010).

Table 2.2 Comparison unit outline

2nd grade compare unit		
Lesson	*Problem*	*Summary or purpose*
1	***Intro comparison with model: blocks/squares*** Jordyn built a tower with 8 blocks and Brendan built a tower with 5 blocks. How much taller is Jordyn's tower? Justify your answer with a model to convince the class.	When comparing two numbers the term "difference" refers to how many more or how many fewer one quantity is than the other.
2	***Introduce number line to show comparison as well as build equation to match the problem together – Difference Unknown*** Bella has 11 cents and Reef has 25 cents. How much more money would Bella need to have as much money as Reef?	A number line as a model helps us see the problem.
3	***Introduce bar model to show comparison*** On Saturday the temperature was 48 degrees and on Sunday it was 89 degrees. How many fewer degrees was Saturday than Sunday?	A bar model or number line are useful models for understanding the problem.
4	***Larger unknown*** Jeanne has 124 more Legos than Nathaniel. Nathaniel has 135 Legos. How many Legos does Jeanne have? Justify your answer with a model and solve.	Labels and models help us to make sense of problems. It helps you see what is missing and what is NOT missing. "More" can mean to add.
5	***Compare smaller unknown*** Oskar has 15 cents fewer than Nellie. Nellie has 31 cents. How much money does Oskar have?	"Fewer" can mean to subtract.
6	***Compare larger unknown*** Dylan has 16 more erasers than Alina. Alina has 37 erasers. How many erasers does Dylan have?	Labeled equations are models.
7 Research Lesson	**Compare "fewer" unknown** **Preston has 12 more pennies than Sophia. Preston has 38 pennies.** **How many pennies does Sophia have?**	**Models can help to determine the operation needed because sometimes "more" means to subtract.**

(*Continued*)

Table 2.2 (Continued)

2nd grade compare unit		
Lesson	Problem	Summary or purpose
8	Luke has 12 cents fewer than Evie. Luke has 37 cents. How much money does Evie have?	"Fewer" can mean to add or subtract.
9	2-Step Problems involving comparison (e.g. A pencil costs 59 cents. A pencil costs 20 cents more than a sticker. How much do a sticker and a pencil cost together?)	*TBD*

Lesson plan

Title of the lesson: Solving a Compare Problem with the Word "More" When the Smaller Quantity Is Unknown

Students' school: Daves Avenue Elementary, Los Gatos, California

Student ages: 7–8 years old

Instructor: Rebecca Setziol

Co-authors: Teresa Bonner, Megan Mahoney, Rebecca Setziol, Carla Pescatore, Clare Brightman, Jana Morse

Date: March 31, 2016

Goal of the lesson: Students will use models and labels to successfully reason through a challenging comparison problem featuring the keyword "more" to determine the correct operation.

Learning standard: CCSS Math 2.OA.1: Using addition and subtraction within 100 to solve one- and two-step problems involving situations of adding to, taking from, putting together, taking apart, and *COMPARING* with unknowns in all positions (Common Core State Standards Initiative, 2010).

Flow of the lesson

Posing the problem

After students gathered on the classroom rug, a student was prompted to read the problem aloud from the board:

Preston has 12 more pennies than Sophia. Preston has 38 pennies. How many pennies does Sophia have?

The teacher then instructed her students to close their eyes and imagine the situation as she read the problem aloud once more.

HATSUMON: "Before you return to your desks to get started on the problem, do you understand what you are being asked to do?" (It should be noted that because our focus was on supporting students to make sense of a new problem situation, we intentionally kept the numbers small to ensure that the calculation itself would not interfere with our stated goal.)

Anticipated responses

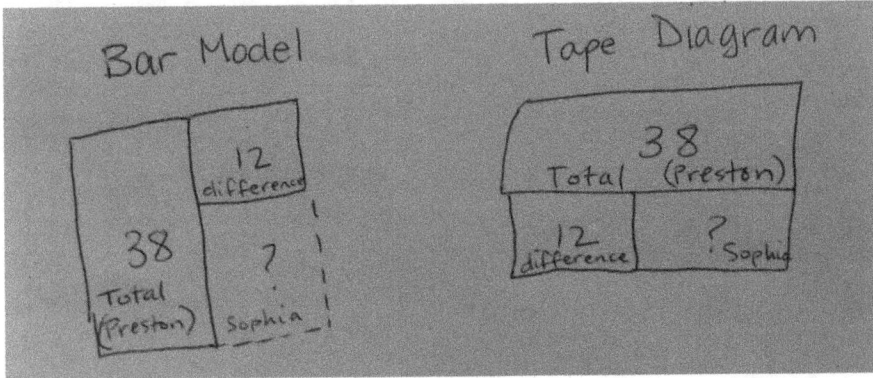

Figure 2.2 Anticipated response 1 (correct)

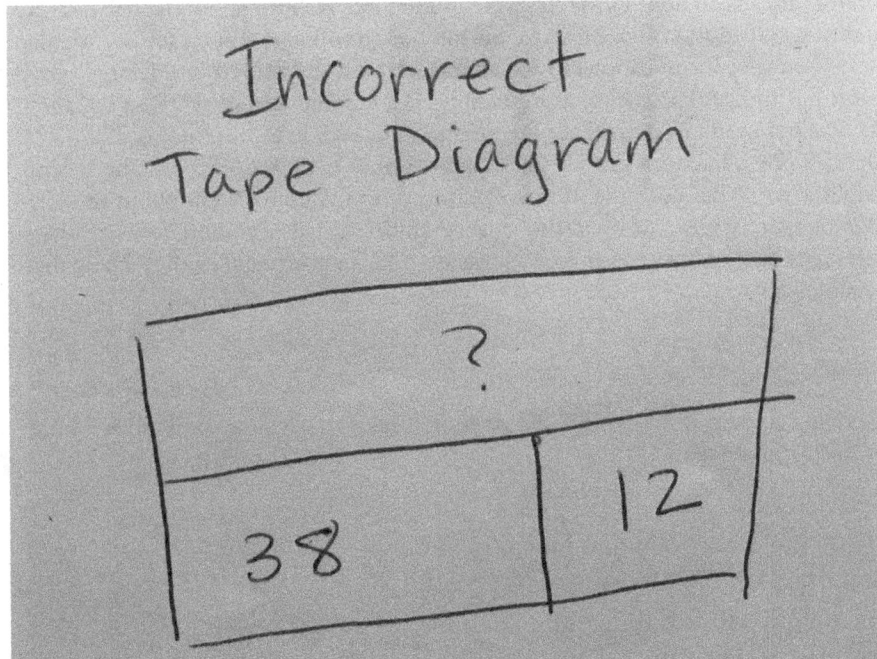

Figure 2.3 Anticipated response 2 (incorrect)

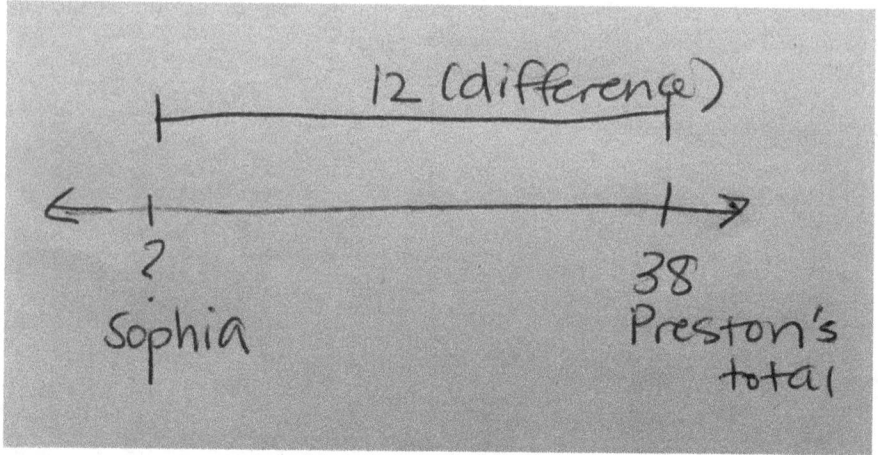

Figure 2.4 Anticipated response 3 (correct)

Comparing and discussing

In our eagerness to encourage and support a productive class discussion, our plan was to set up a controversy based on the anticipated misconception that "more" means to add, by having one student share her method to solve using addition and another student share his method to solve with subtraction. We also discussed specific questions in advance that might help elicit our desired board work during the discussion. For example, if a method was missing an equation, the teacher might ask, "Is there an equation that would represent this diagram?" Or, if there were numbers on the board without contextual labels, the teacher might ask, "What does the 38 represent in your diagram? What about the 12?" Our questions were designed to ensure students had a comprehensive visual reference that could anchor the discussion and support them as they made their arguments.

Summing up

Models can help to determine the operation needed because sometimes "more" means to subtract.

Board plan

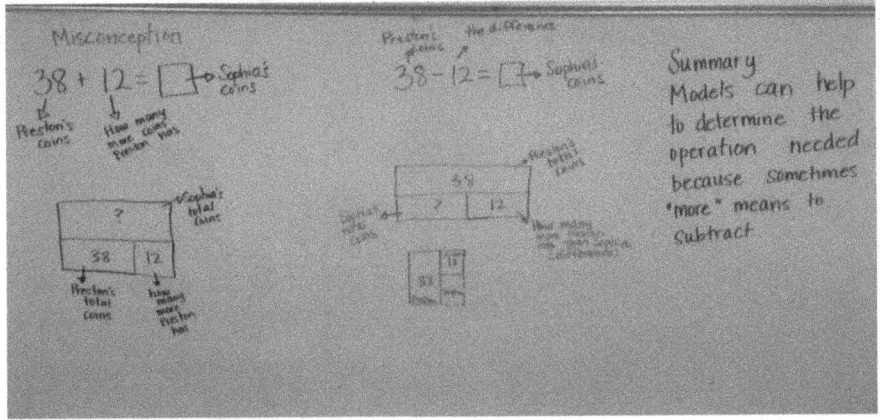

Figure 2.5 What we anticipate

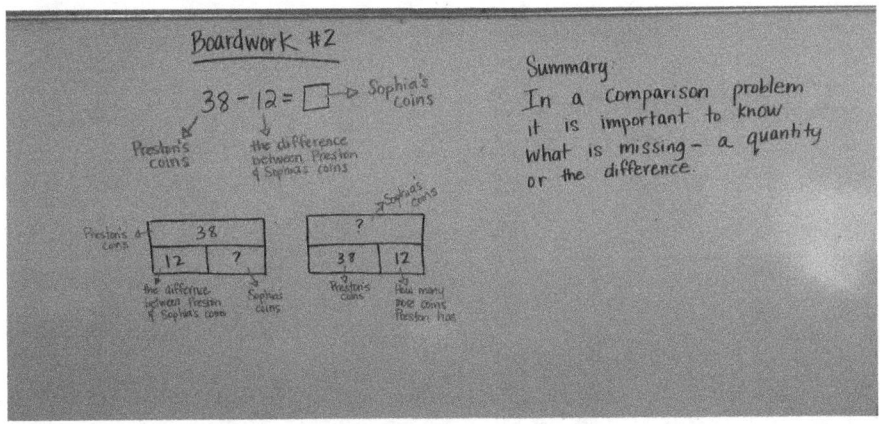

Figure 2.6 Plan B if misconception does not arise

Observations from the lesson

Many students drew models as they worked independently on their ideas (tape diagrams, number lines, equations), though not necessarily in ways we had expected.

Some students used the number line as a tool for solving, not for reasoning (see Figure 2.8 a and b).

In other cases, students identified the operation (whether correct or incorrect) first with an equation, then drew a tape diagram for justification. These students were not yet ready to use a tape diagram as a tool for reasoning (see Figure 2.9 a and b).

Figure 2.7 The board at the conclusion of the lesson

Figure 2.8a The student used a number line to add 12 to 38, then checked the answer using a number line

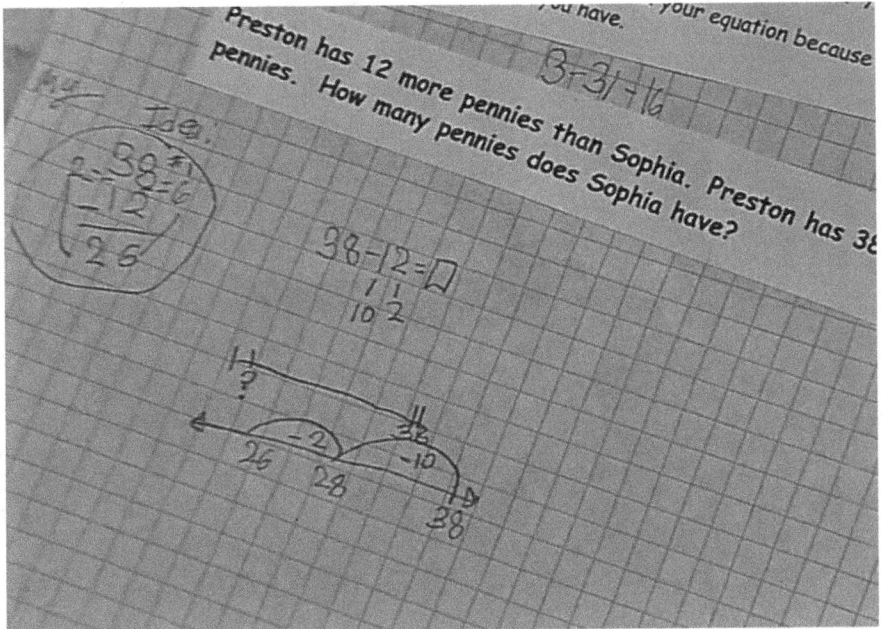

Figure 2.8b The student calculated 38 − 12 with an algorithm

Learning highlight from post-lesson discussion

Because the class was fairly divided on whether to add or subtract as we had anticipated, the teacher was able to capitalize by selecting two students to share their conflicting solutions. After both (a tape diagram model and bar model) were recorded on the board and the quantities were labeled with help from the class, students were encouraged to make their case.

During the post-lesson discussion, we concluded that although students cited the word "more" in arguing for addition, the summary statement that included "sometimes 'more' equals 'less'" did not reflect the actual student learning. The focus of the class discussion centered around two male students (considered strong in math by their peers) who refused to accept that it was a subtraction problem and continued to cite the use of "more" to support their argument for addition. It was not until one student recalled his prior learning and recognized Preston's quantity as equivalent to the "whole" before he was able to see that 12 represented the difference and should be subtracted from Preston's 38 pennies to determine Sophia's amount. According to Professor Tad Watanabe of Kennesaw State University:

> Japanese teachers often talk about expanding the meaning of operation. So, they might start with the "take-away" meaning for subtraction, but they want students to expand the meaning to include PPW-part-unknown,

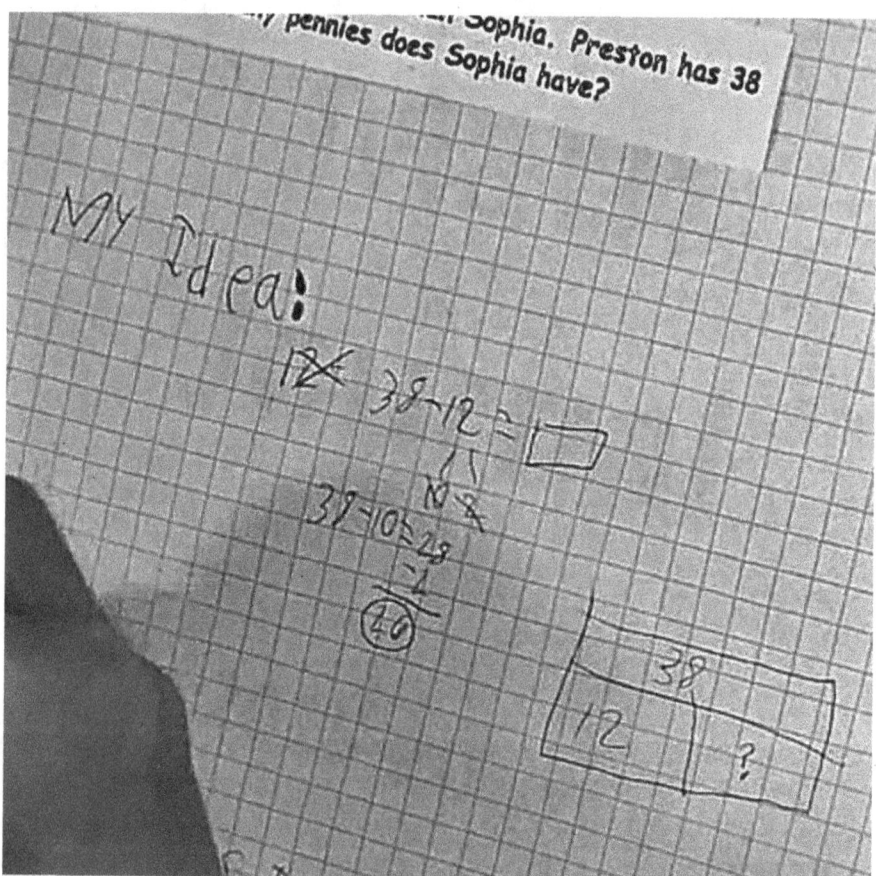

Figure 2.9a The student *correctly* identifies the operation in an equation and justifies his reasoning with a tape diagram

comparison-difference-unknown, and other types by connecting the new situations to what students learned – and developing the disposition to connect the new/novel situation to prior learning.

(Watanabe, 2016)

In retrospect, our summary should have focused on the usefulness of connecting prior understanding to new situations because it was this reason alone that accounted for the student's breakthrough in understanding and brought the whole class together in agreement.

Figure 2.9b The student *incorrectly* identifies the operation in an equation and justifies her reasoning with a tape diagram

What I learned

As is always the case after a Lesson Study cycle, the learning is never ending and covers many aspects of the teaching and learning process. For this particular Lesson Study cycle, however, what stood out for me most was the importance of using models as tools for reasoning.

First, I learned the distinction between a model used for reasoning versus one used for solving. We had anticipated that students might use a number line model to represent the subtraction situation, but instead the number line was used to calculate the solution. And though Professor Watanabe had pointed this possibility out before our research lesson, the point was overlooked until after the Lesson Study cycle was complete.

Secondly, I learned *how* it was possible for a model to be used as a tool for reasoning. The numbers displayed in the tape diagrams were removed from the problem and therefore de-contextualized. This allowed the student with the eventual epiphany to think about the three quantities in more familiar terms (a part, a part, and a whole). Despite the fact that the problem types are different (Compare vs. Part-Part-Whole), the equations for each are mathematically equivalent and can be viewed as such when modeled in a tape diagram.

Impact on my work

For students to gain deeper conceptual understanding, it is imperative that they connect what has already been learned and apply it to new problem situations. It is for this reason that I encourage teachers to not only study the standards at their grade level, but to understand the standards that come before and follow after. This way, teachers can be intentional about the order in which their lessons are presented, creating opportunities for students to make connections to multiple mathematical ideas and concepts simultaneously.

I also learned the value of researching common misconceptions. Recognizing that students came to grade 2 with a limited understanding of the meaning of addition and subtraction was key in creating a need for us to investigate problem structures and formulate a plan that would connect their narrow definition to new situations.

References

Bonner, T., Mahoney, M., Setziol, R., Pescatore, C., Brightman, C., & Morse, J. (2016). *Comparative Subtraction. Research lesson proposal.* https://LSAlliance.org/Lessons

Carpenter, T. P. (1999). *Children's mathematics: Cognitively guided instruction* (1st ed.). Heinemann.

Common Core State Standards Initiative. (2010). *Common Core State Standards for Mathematics.* http://www.corestandards.org/Math/

Watanabe, T. (2016). *Commentary.* In Bonner, Mahoney, Setziol, Pescatore, Brightman, & Morse, *Comparative Subtraction. Research lesson proposal.* https://LSAlliance.org/Lessons

Expanding the meaning of multiplication

Understanding multiplication of a whole number by a decimal, grade 5 (10- and 11-year-old) students

Joshua Lerner

When asked to describe the meaning of multiplication, many students might explain how to use repeated addition to find a total amount. Indeed, in grade 2, we help students see that same sized sets can be added using repeated addition, and, in grade 3, we formalize this concept as an operation called multiplication. However, while this new operation feels exciting for students, we as teachers secretly know that their "equal groups" or "repeated addition" understanding of multiplication is still quite limited. Within just a few years, students will begin to understand multiplication in a much more powerful way: as a concept of proportional relationships. From a research lesson our team conducted in 2020, I discovered that learning to multiply by a decimal is an opportunity for students to expand their understanding of multiplication toward proportionality.

According to the Common Core State Standards, students in grade 5 learn to "add, subtract, multiply and divide decimals" (Common Core State Standards Initiative, 2010). But with multiplication, it makes a difference conceptually if the decimal is the multiplicand or the multiplier. For example, multiplying a decimal by a whole number is relatively straightforward. Let's say I have a set of books with widths of .3 cm. How tall is a stack of six of these books? Six sets of 0.3 can be thought of as a case of repeated addition, albeit with the equal size quantities as a non-whole number. But what about when we multiply a whole number by a decimal – when the decimal is the multiplier? Take the following: I have a length of rope that is one meter long and weighs 80 grams. What will be the weight of 2.3 meters of this rope? Of course, if students were asked to find the weight of 2 or 3 meters of this rope, the calculation would be simple. But if students have only understood multiplication as repeated addition, then how will they think of multiplying by 2.3? Perhaps they will wonder, does this situation even count as multiplication?

Of course, this *is* multiplication. In fact, it is an early opportunity for students to think of multiplication as a representation of a proportional relationship. In our research lesson, which used this problem as its central task, our goal was for students to understand that multiplication can still be used with problems such as this one. We used a double number line to help students make sense of this new learning. A second goal of our lesson was for students to use what they had learned previously about multiplying (decimal) by (whole number) to see that the calculation process is the same for (whole number) by (decimal). The effects of this lesson on student learning were somewhat mixed, partly as a result of the conflation of these two learning goals. But, as shown next, there was a lot for us to learn as teachers!

DOI: 10.4324/9781003230915-8

Lesson plan (shortened)

Title of the lesson: Multiplying a Whole Number by a Decimal
Students' school: Helen C. Peirce School of International Studies
Student ages: 10–11 years
Instructor: Joshua Lerner
Co-authors: Yael Berenson-White, James Callerstrom, Vivian Leventis
Date: January 14, 2020
Goal of the lesson:

- Students understand that multiplication can still be used even if the multiplier is a decimal.
- Students can solve problems involving a whole number multiplied by a decimal number by using what they already know, including the idea of relative size (*10 times* and $\frac{1}{10}$ *of* relationships) and properties of multiplication.

Learning standard: 5.NBT.B.7. Add, subtract, multiply, and divide decimals to hundredths, using concrete models or drawings and strategies based on place value, properties of operations, and/or the relationship between addition and subtraction; relate the strategy to a written method and explain the reason used.

Flow of the lesson

Introduction

Present a 2.3 m section of rope, without revealing its length, along a line that is marked with tick marks at 1, 2, and 3 m. Explain that 1 m weighs 80 grams. We want to know how much this rope weighs. Lead students through considering how much 2 m and 3 m would weigh.

Posing the problem

Reveal today's problem: "1 meter of rope weighs 80 g. I bought 2.3 m of the rope. How much does it weigh?" Ask students to discuss and come up with a number sentence to represent the situation. Help students confirm that this is multiplication, and can be written as $80 \times 2.3 = ?$[1] Ask: "How can we use what we already know to multiply by a decimal?" Students begin working independently.

Anticipated responses

R1. Student does not know how to get started, perhaps unsure of whether addition or multiplication should be used and why.

R2. $80 \times 2.3 = 184$. Student sets up a vertical algorithm and calculates correctly. However, the student may not understand the parts of their calculation (240 and 1600, before placing the decimal point) or what it means to have 2.3 of something.

R3. The student multiplies 2.3 by ten, and then divides the quotient by 10:

 – The weight of 23 m would be 1840 g.
 – The real weight would have to be $\frac{1}{10}$ of 1840 g, so the answer is 184 g.

R4. The student follows these steps, or attempts to and cannot find the weight of .3 m rope:

 – $80 \times 2 = 160$
 – Since each .1 m would weigh 8 g, the weight of .3 m would weigh 24 g.
 – $160 + 24 = 184$.

R5. The student follows these steps:

 – The weight of .1 m is 8 g.
 – There are 23 0.1 m units, so $8 \times 23 = 184$.

Comparing and discussing

Students present responses in the following sequence, R4, R3 or R5, then R2, with comparison and discussion across responses.

For R4, encourage students to discuss why it was useful to find the weight for 2 m of the rope and how the student was able to find the weight of the additional .3 m of rope. Help students see the importance of finding that the weight of .1 m is 8 g.

Connect this last aspect of R4 to R5. Help students see this connection by drawing a tick mark at .1 m and labeling it as 8 g. Ask, "How can we use this tick mark to find the total weight for 2.3 m?"

Connect R3 and R2 by showing that both processes involve thinking of 2.3 as a whole number by first multiplying by 10 and then multiplying the final product by $\frac{1}{10}$. $80 \times 23 = 1,840$. We take $\frac{1}{10}$ of 1,840 in order to find the weight of the 2.3 m rope = 184 g.

Summing up

Help students summarize new learning through a statement such as the following: "Even when we are multiplying by a decimal, we can solve the problem by using what we already know – such as *10 times* and $\frac{1}{10}$ *of* relationships." Students then reflect on their own takeaways from the lesson.

Board plan

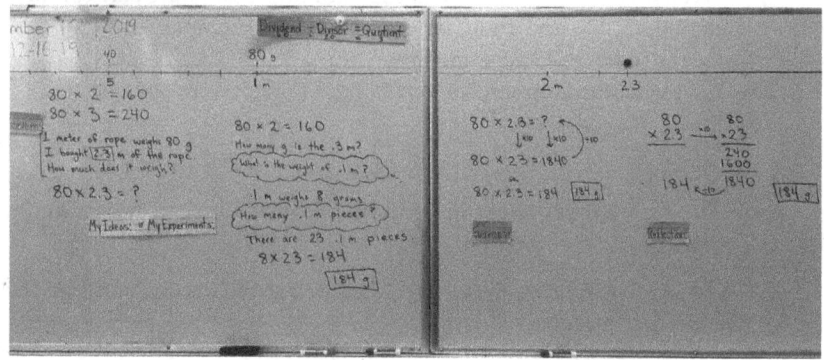

Figure 2.10 The board plan for the lesson

Observations from the lesson

To begin the lesson, we introduced a length of rope along a number line indicating the length of 1 meter, 2 meters, and 3 meters. After telling the students that 1 m of rope weighs 80 grams, we asked the students to estimate the weight of the length of rope on the board. As the teacher leading this part of the lesson, I asked the students to first consider how much 2 m and 3 m of rope would weigh and how we would represent those calculations. As they responded, I made their thinking visible on the board:

80 × 2 = 160, 2 m of rope weighs 160 g
80 × 3 = 240, 3 m of rope weighs 240 g

When I finally revealed the length of rope on the board and asked students to consider how they would find its weight, their contributions came quite naturally. "We can use multiplication!" "Just use 80 × 2.3!" But what did these immediate responses reveal about student understanding? Did students find the question straightforward because they were simply following the multiplication pattern that I had set up in the introduction to the problem? I had a sneaking feeling that students had missed out on the foremost tension of the lesson: the question of whether we can even multiply by a decimal number such as 2.3 to begin with! And now that students had, seemingly by rote, decided they could express this situation using multiplication, how could I help them take an interest in the special nature of what it means to multiply by a decimal number?

Within a few minutes, the students were back at their seats working independently in their math journals. As we circulated between desks, we observed that most students made use of the strategies they had previously learned for multiplying a decimal by a whole number. It was then that I developed a new feeling of trepidation. They didn't seem to view this multiplication situation as any different than the ones they had solved before. Of course, multiplication is commutative, and so students could very well use previously learned strategies to solve this new type of problem correctly. But what was the point of having lesson goals about making sense of a decimal multiplier, if it appeared that students were carrying out their mathematical work without seeming to notice they were dealing with a decimal multiplier at all! As I observed students and planned for discussion, I kept coming back to the following question: what was the connection between our first lesson goal – learning that multiplication can still be used when multiplying by a decimal – and our second goal – learning how to calculate such a multiplication problem?

In the end, the class did have a rich mathematical discussion that focused on the connection between R2 and R4 (as listed earlier). Students interpreted the work of one of their peers, who multiplied 80 x 2 = 160, but struggled to find the weight of the remaining .3 m of rope. Through discussion, students came to consensus that since .1 m would only weigh 8 g, the remaining .3 m would then weigh 24 g. They then compared this process to the standard algorithm, presented by another classmate, in which they could see the numbers 160 and 24 in plain sight. The lesson ended neatly, but, as is so common with research lessons, I was left wondering to what extent our learning goals had truly been met.

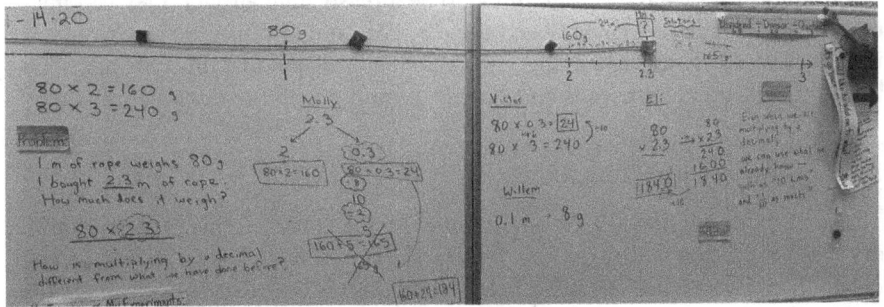

Figure 2.11 The board at the end of the research lesson

What I learned

One of the gifts of Lesson Study is the opportunity to learn by doing – to be "in practice" with your fellow teachers, and then to be able to sit and talk with them and gain insight from a shared experience. From the post-lesson discussion and final comments of our research lesson, I was indeed able to resolve some of the tensions I had experienced while teaching the lesson.

In his final comments, Akihiko Takahashi focused on the stage in which we introduced the problem and helped students make sense of it. For him, this was the "heart" of the learning for today's lesson. He agreed with other observers that we had missed an important opportunity. To help students understand what it means to multiply by a decimal, we had to first consider for ourselves how proportional thinking is related to, yet different from, equal groups thinking.

What if we had asked the question: "If a 1 m rope weighs 80 g, what is the weight of the rope that is 2.3 times as long as this rope?" In this "times as much" situation, the role of multiplication is continuous. Students could think of this new rope as stretching, or scaling, continuously to a new length that is 2.3 times the length of the first. The resulting weight would be the number that corresponds to that new number of meters, just as 80 g corresponds to 1 m. The double number line would help students visualize this relationship. Students would then realize that multiplying by 80 could help them find the weight of a rope of any new length, even a non-whole-number length. Moving from equal groups thinking to proportional reasoning, students would begin to develop a fuller understanding of the meaning of multiplication.

The post-lesson discussion also helped me make sense of my question about the two goals and how they related. I learned that whenever we ask students to expand their definition of a mathematical concept, it is helpful to frame the lesson as a test case using what they have previously learned. In this case, we could have used the first part of the lesson as a hypothesis ("I wonder if this can be represented as 80 × 2.3") and then the calculation phase as a test of our idea ("Let's see if we can use what we know about multiplying with decimals to solve this new kind of multiplication problem"). Then, when students successfully use the multiplication algorithm to find what they determine to be a reasonable answer, they can interpret their conjecture as having been correct. They might finish the lesson with an insight such as, "It seemed strange at first, but it turns out that this type of a situation with a decimal number is an example of multiplication after all!"

Impacts on my own teaching

The next morning, I was back on the rug with the students, this time with just a dry erase marker and a rubber band. I was there to teach a short "re-engagement lesson," applying what our team had learned during yesterday's post-lesson discussion (Inside Mathematics, 2021). I challenged the students to consider again how multiplying by a decimal is different from what they had done before.

I first led them through a guided demonstration, in which I stretched the rubber band along a number line until it was two times its original length and three times its original length. I then stretched the rubber band to 2.3 times its original length – and made a big deal of it! I asked students to consider whether multiplication could still be used to describe this "times as much" relationship, even though our new length fell in between the whole numbers of 2 and 3.

We then applied this model to yesterday's problem. We discussed explicitly that, by using "times as much" thinking, we could use multiplication to find the weight of any length of rope, even if it were not a whole number. This meant we could now multiply by decimal numbers! Of course, I was careful to note that, in yesterday's problem, we were dealing with a new, longer rope – not a rope that

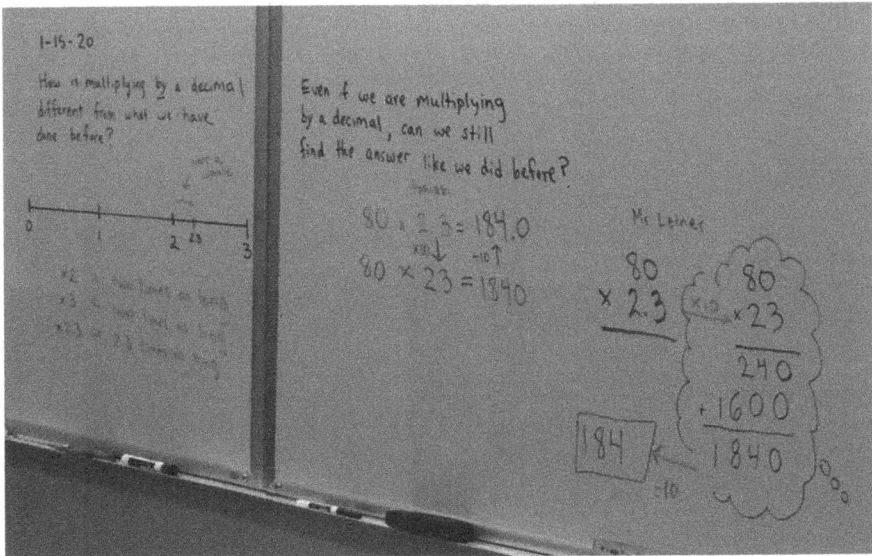

Figure 2.12 The board at the end of the re-engagement lesson

was itself being stretched – that would thus be a new, greater weight. It felt good to know that I was helping students better meet the lesson's first goal as a result of what I had learned from the post-lesson discussion.

I have also recently applied my learning from this research lesson in my work with a seventh grade class. Much of the grade 7 curriculum focuses on proportional reasoning. A common problem goes something like: "A recipe calls for 1 tablespoon of baking powder for every 3 cups of flour. If 2.5 tablespoons of baking powder are used, how many cups of flour will be needed?" I now use double number lines and concrete models that demonstrate the concept of scaling in order to help students think multiplicatively. They begin to approach a problem like this by thinking, "If the number of tablespoons is now 2.5 times as much, I can multiply 3 × 2.5 in order to find the amount of flour that is also 2.5 times as much."

As I reflect on my own professional growth from this research lesson, I see how it all comes together. As I taught the lesson, I felt the type of intellectual tension that comes with knowing you are on the cusp of learning something new and important. During the post-lesson discussion and final comments, I was able to synthesize critical insights in a lasting way because they were based directly on my own teaching and the planning of our team. Moving forward, I was able to better understand the developmental trajectory of an important topic and apply it to my practice across multiple grade levels. This is the type of long-term professional learning that helps us as teachers grow professionally over our careers. And we have the practice of Lesson Study to thank for that growth.

Note

1 As a result of our use of Japanese curricular materials, students at Peirce are used to writing multiplication sentences as they are written in East Asian countries, with the multiplicand first, followed by the multiplier. In this case, $80 \cdot 2.3$ would be read as "80 multiplied by 2.3."

References

Common Core State Standards Initiative. (2010). *Common Core State Standards for Mathematics*. www.corestandards.org/Math/

Inside Mathematics. (2021). *Formative re-engaging lessons*. www.insidemathematics.org/classroom-videos/formative-re-engaging-lessons

Conceptualizing equivalent expressions through problem-solving

Combining like terms, for grade 7 (12- to 13-year-old) students

Aaron Bingea

In 6th through 8th grade, students are tasked with developing essential understandings and skills necessary for future high school coursework in algebra. Ideally, middle schoolers will take their knowledge of numbers and operations from earlier grades and begin to reason about abstract algebraic expressions. In my experience, this is a fragile and challenging moment in a student's mathematical journey. When moving through our traditional curriculum, students experience significant trouble when presented with pages of decontextualized coefficients, variables, and constants. Working with algebraic expressions can become off-putting to the point where some students will develop a reluctant and negative attitude towards the entire subject of algebra. As a teacher of these foundational algebra units, I have shied away from addressing these issues head-on. To ease my students' discomfort, I often found myself giving students lifelines in the form of tricks and easy-to-remember procedures when learning concepts such as simplifying expressions or solving equations. These lessons could be summed up with the phrase, "Ok, when you see this, do this." When resorting to such measures, I knew that I was robbing my students of an opportunity to do meaningful math in the name of getting through the curriculum. Students in my class would become competent at following rules to find an equivalent expression but would not express the underlying mathematical principles useful for future mathematical topics.

Our middle school research team, composed of fourth–eighth grade teachers, met to choose a research lesson topic; we found everyday struggles in teaching units that required students to reason with algebraic expressions. It proved to be a valuable topic to address across all grade levels and had implications in every team member's curriculum and teaching practices. We then examined our curriculum maps and identified the standard that was frequently taught void of meaningful problem-solving and time for sense-making, combining like terms.

We felt that how our school curriculum treated this topic, and how we approached this topic, fell short of our ultimate goal of fostering conceptual understanding through problem-solving that would lead to future access to algebraic concepts. Most lessons involved students seeing expressions and working to generate equivalent expressions by applying the distributive property and collecting like terms. When examining summative assessments from previous years, it

DOI: 10.4324/9781003230915-9

was evident that students could carry out the procedures of combining like terms. Still, it provided little or no justification in their work when prompted. We concluded that our students were missing opportunities to develop an argument for why and combine like terms to generate equivalent expressions.

At the time of this Lesson Study cycle, our school's research theme was to develop the mathematical practice of constructing viable arguments and critiquing the reasoning of others. So, in addition to creating problems that would provide students with sound mathematical justifications for why we can combine like terms, we aimed to develop lessons that would get students talking, listening, and debating about the math they were engaging with. Throughout the research phase, our team examined the treatment of combining like terms across several different curriculums. We learned that these curricula almost exclusively dealt with these concepts in the abstract, void of any contextualized problem-solving, and gave students very little prompting and time to grapple with why simplified expressions are equivalent. To promote the mathematical practice of constructing arguments and critiquing the reasoning of others, we needed to include problem contexts that we knew would be relevant and accessible to our students and use the teaching through the problem-solving model to give them space for these conversations.

In the research lesson, we wanted students to do two things mathematically. 1) Understand why we can combine like terms by reasoning with expressions derived from a real-world problem. 2) See that you can combine terms by factoring out common variables. These are tricky concepts for seventh graders to grasp and ones that we previously had not dedicated any time in this unit in past years.

To further illustrate the issue we were trying to address, consider this focus problem from our textbook:

Write each expression with fewer terms. Show your work or explain your reasoning.

1    $10x - 2x$
2    $10x - 3y + 2x$

(Illustrative Mathematics, 2017)

Typically my students would answer these two problems by combining the terms that look similar and attending to the coefficients. A student might go so far as to argue that 10 x's subtracted by 2 x's makes 8 x's. This logic holds but fails to prove why we shouldn't combine $10x$ and $-3y$ in item number 2. So when asked to justify why $10x - 3y + 2x$ is equivalent to $12x - 3y$, students will generally respond with some iteration of 'we can't combine x's and y's.' We wanted students to build on their previous understandings of the distributive property and common factors and notice that $10x + 2x = x(10 + 2)$. Students should see the common variable in a term as a common factor. We hypothesized that given an accessible context, students would be able to reach this understanding.

Our unit research lesson in this unit set out to remedy this. We deliberated on problem contexts and decided to use the scenario of filling up a small pool with different hoses and asking students to find how many gallons of water were in the pool after different amounts of elapsed time. Students would compare and discuss their expressions and ultimately figure how much water would be in the pool after 'x' minutes. Then different student responses would be highlighted for students to grapple with their equivalency.

Lesson plan (shortened)

Title of the lesson: Grade 7 Combining Like Terms
Students' school: Brentano Math and Science Academy, Chicago, IL
Student ages: 12–13
Instructor: Aaron Bingea
Co-authors: Erendira Alcantara, Aaron Bingea, Cassie Kornblau, Martin Lenthe
Date: May 2018
Goals of the lesson:

a) Students will understand why like terms can be combined to create an equivalent expression.
b) Students will understand why a term with a variable cannot be combined with a constant.

Learning standards

CCSS Math 7.EE.A1.b Apply properties of operations as strategies to add, subtract, factor, and expand linear expressions with rational coefficients.

CCSS Math 6.EE.A.3 Apply the properties of operations to generate equivalent expressions. *For example, apply the distributive property to the expression $3 (2 + x)$ to produce the equivalent expression $6 + 3x$; apply the distributive property to the expression $24x + 18y$ to produce the equivalent expression $6 (4x + 3y)$; apply properties of operations to $y + y + y$ to produce the equivalent expression $3y$.*

(Common Core State Standards Initiative, 2010)

Flow of the lesson

Introduction

This lesson will give students several opportunities to develop an argument for combining like terms. The central problem involves a pool being filled up with water by multiple hoses that fill at different rates. Students will be asked to explain how they calculated the amount of water in the pool after a given number of minutes and eventually pushed to generating an expression to model the amount of water in the pool after x minutes. We expect students will naturally start to combine like terms after multiple iterations of this problem and will be able to create an argument as to why certain terms can or cannot be combined by using the problem context and the distributive property to justify.

Posing problem 1

Display problem on board:

PROBLEM: A swimming pool starts with 3 gallons of water. Two different hoses are turned on and begin filling up the pool. The first hose fills up the pool at a rate of 2 gallons per minute. The second hose fills up the pool at a rate of 4 gallons per minute.
How much water is in the pool after 5 minutes?

TEACHER: Take some time to solve this in your notebooks. Show the calculations you used to find the answer. Draw a picture if it would help.

Students work independently for 3 minutes.

Anticipated responses

R1: $3 + 2(5) + 4(5) = 33$
R2: $3 + 6(5) = 33$
R3: $9(5) = 45$ (misconception)

Discussion

Highlight R1: Ask whole class to discuss what every term means.
Highlight R2: Ask whole class to discuss what this student did.
Misconception: Combining the starting value

Ask the whole class to explain what the student who came up with R3 may have been thinking. Do you agree?

Posing problem 2

How much water is in the pool after 7 minutes?

TEACHER: Take some time to solve this in your notebooks. Show the calculations you used to find the answer. Draw a picture if it would help.

Students work independently for 3 minutes.

Anticipated responses

R4: $3 + 2(7) + 4(7) = 45$
R5: $3 + 6(7) = 45$
R6: $9(7) = 63$ (misconception)

Ask the whole class to discuss. What is the same/different about your work on these two problems?

Discussion

Highlight that only the number of minutes changes.

ASK STUDENTS: How much water is in the pool after x minutes? Generate an expression to represent the number of gallons in the pool after x minutes.

Students independently generate expressions to represent the situation.

Anticipated responses

R7: $3 + 2x + 4x$
R8: $3 + 6x$
R9: $9x$

Prompt students to turn and talk: Which expression(s) do you agree with and why?
Questions to discuss whole group:

Does expression R7 match this situation? Solidify this first so that R8 and R9 can be discussed in comparison to A.
After student responses, teacher will present
$3 + ()x$ and ask where the '6' came from.

Is expression B equivalent to expression A? How do we know?
Is expression C equivalent to B/A? How do we know?
Why does expression C not work? Why can't we just combine all the numbers?

Posing problem 3

A swimming pool starts with 40 gallons of water. Two hoses are turned on and begin filling up the pool. Hose A fills up the pool at a rate of 37 gallons per minute. Hose B fills up the pool at a rate of 13 gallons per minute. There is also a leak in the pool, and it LOSES 10 gallons per minute.
How much water is in the pool after x minutes?

TEACHER: Take some time to solve this in your notebooks. Show the calculations you used to find the answer. Draw a picture if it would help.

Students work independently for 3 minutes.
Present the anticipated responses:

R10: $40 + 37x + 13x - 10x$
R11: $40 + 50x - 10x$
R12: $40 + 60x$ (misconception)
R13: $40 + 40x$
R15: $40 + x(37 + 13 - 10)$ or $40 + (37 + 13 - 10)x$
R16: $80x$ (misconception)

Discussion

Clear up any misconceptions, by having students look for any errors. Students will address these errors by talking to their table partners and then clearing.

How do we know R10 and R11 are equivalent?
Why are R10 and R13 equivalent?
Using your same argument: Why can't we combine all of the terms in the expression? Why can't we combine the 40 with the 37 and the 13?

Summing up

Students respond to one last prompt:

$4 + 5x + 2x$

Make an equivalent expression with fewer terms. Explain why we are allowed to do this.

Observations from the lesson

Our team was satisfied with the students' responses and discussions that occurred during the lesson. The lesson proved to elicit the desired misconceptions and varying strategies to combine like terms. The majority of students used the context of filling up a pool with water to justify why specific terms could or couldn't be combined. We also felt that students were using each other's arguments to refine their own during turn and talks and in whole-group discussions. For instance, when one student referred offered their expression for problem 3, they stated that they used Andre's method from the last problem. Throughout the lesson, students referred to an idea of another classmate to make a point or explain their answer.

Overall we thought the lesson fell short in meeting our most advanced goal of having students justify combining like terms using the distributive property. Many students were able to state why we can combine the rates of each hose because they were both being multiplied by minutes. However, we did not see any evidence that students abstracted this idea to the point where students combined like terms by factoring out the variable. In the post-lesson discussion, several observers saw evidence of the distributive property used by students, while other felt that overall the class seemed to spend most of their time thinking about why the terms with coefficients couldn't be combined with the constants. After reflecting on this fact and the post-lesson discussion where this was debated, we concluded that this was an issue with the lesson goal and not the lesson itself. Because this was the students' first experience with combining like terms, our goal should have focused more on using the context to justify why we can combine like terms and why it can be helpful. We felt this goal was largely achieved and properly set students up to address the more abstract justification using the distributive property in the next lesson. We also concluded that in order for students to apply the distributive property to concepts like these, we need to make sure it is given more weight and time for conceptual development in earlier grades.

In our research, we could not find a curriculum that treated this topic first with a concrete problem-solving context. Instead, most units taught combining like terms with algebraic expressions devoid of any context. This lesson proved that students benefited from making connections to the different terms with a relevant real-world situation and allowed most students to construct a sound mathematical justification for combining like terms. We have traditionally taught this skill briefly in the abstract without giving students time to develop a robust rationale for the skill.

What I learned

This research lesson highlighted the essential nature of keeping problem-solving at the center of instruction. In the traditional algebra curriculum, problem-solving opportunities are often relegated to the end of the lesson – something extra for students to work on. I have been guilty of initially and primarily tasking students with solving algebraic exercises in the abstract and then tasking them to apply it in a real-world context. In this lesson, we led with a familiar context that allowed students to ground their understanding of the different terms with concrete quantities. This instructional decision provided a more accessible conversation when we shifted focus to abstract algebraic terms. Although we did not reach our initial goal of students justifying combining like terms using the distributive property, in the following lessons students were much better positioned to grapple with this abstract way of justification due to their experience with this lesson. Past treatments of this topic would have only afforded students who could intuit the justification behind combining like terms to do meaningful mathematics. This lesson reinforced the need to offer relevant and accessible problem-solving opportunities as a vehicle to addressing abstract math content.

Impacts on my own teaching

The success of this lesson in this regard has led our team to consider more algebraic skills and concepts that could be taught first with problem contexts so that students can develop mathematical arguments behind algebraic moves before applying them in a purely abstract context. Some units where we have increased opportunities for problem-solving include solving equations in seventh grade and systems of equations in eighth. Slowing down and allowing time for students to grapple with relevant contexts at the beginning of units has not only benefited my students' conceptual understanding of algebraic skills, but also minimized the frustration and burden of developing procedural fluency.

References

Common Core State Standards Initiative. (2010). *Common Core State Standards for Mathematics*. www.corestandards.org/Math/

Illustrative Mathematics. (2017). *IM 6–8 Math: Grade 7 Mathematics*. https://access.openupresources.org/curricula/our6-8math/en/grade-7/teacher.html

Expanding my conception of place value

Three-digit subtraction with regrouping, for grade 3 (8- and 9-year-old) students

Thomas McDougal

In 2016, I had the opportunity to observe a research lesson with grade 3 students that focused on the challenge of subtraction with regrouping. The specific task in the lesson was to calculate $402 - 175$ using the standard algorithm. I was eager to observe this lesson, since I had seen how grade 2 and grade 3 students sometimes struggle to reliably calculate multi-digit subtraction with regrouping. If there is regrouping required in only one place, they typically have no difficulty. But when there is regrouping in two adjacent places, errors become common, and regrouping from a place with a zero, as would be required by this task, is especially troublesome. From observing this research lesson, and from the final comments at the end of the discussion, I gained not only a new idea about how to teach regrouping, but a new way to think about place value.

When I was a school math coordinator, the teachers and I tried to build students' conceptual understanding of subtraction with regrouping by using pictures or base-10 (Dienes) blocks, with only partial success. We discovered some limitations with using base-10 blocks. First, they don't allow actual decomposition, such as breaking apart a ten into 10 ones – students have to "trade" a block for 10 smaller blocks. Second, in contrast to addition, where one can represent both addends simultaneously, there is not an obvious way to represent the subtrahend – students have to keep it in memory, which makes them prone to lose track of what they are supposed to do. And finally, when they use base-10 blocks, students are inclined to begin with the larger blocks. These limitations made it challenging for students to understand what they would do with base-10 blocks related to the standard right-to-left written algorithm (see also Tondevold, n.d.).

The authors of the research lesson described their students having difficulties similar to those I had observed:

> As a team we have noticed that incoming third graders struggle with regrouping especially when it comes to subtraction. Students can subtract whenever there is no need to regroup, but with more complex problems they lack a foundational understanding of what regrouping is and why it is necessary. We decided to . . . focus on subtracting across one zero for the research lesson. Subtraction across zeros is challenging and regrouping can be difficult

DOI: 10.4324/9781003230915-10

to comprehend, thus we have decided to focus on a problem that requires regrouping two times.

In lessons prior to the research lesson, their students solved problems requiring regrouping from both the tens and hundreds places. They had even calculated problems such as $540 - 286$. But this was the first time they were encountering a zero in the tens place when they needed to regroup from there. I was eager to see how their lesson would approach this problem.

Lesson plan (shortened[1])

> **Title of the lesson:** "Let's think about how to calculate 402−175"
> **Students' school:** Dr. Jorge Prieto Math & Science Academy, Chicago, Illinois
> **Grade:** Grade 3
> **Instructor:** Judy Villaseñor
> **Co-authors:** Jennie Vazquez, Dena Kelly, Ashley Schoenecker, Andrew Friesema
> **Date:** November 4, 2016
> **Goal of the Lesson:** The student understands that in order to calculate $402 - 175$ they need to regroup ten ones from the tens place in order to calculate the ones place. Since there are no tens to be regrouped, students have to rely on their previous experience regrouping from the hundreds place to regroup 1 hundred from the hundreds place into 10 tens, and then regroup one of those tens into 10 ones.
> **Learning standard:** Fluently add and subtract within 1000 using strategies and algorithms based on place value, properties of operations, and/or the relationship between addition and subtraction (3.NBT.A.2) (Common Core State Standards Initiative, 2010).

Flow of the lesson

Introduction

Students think about how to calculate $492 - 175$. Brief discussion about regrouping 10 ones from the 9 and then subtracting $12 - 5$ (ones). Students solve and agree on the answer: 317.

Posing the problem

Change $492 - 175$ to $402 - 175$. Students discuss how this is different from the opening problem.

> *Hatsumon*: "Let's think about how to solve $402 - 175$."
> Students work independently.

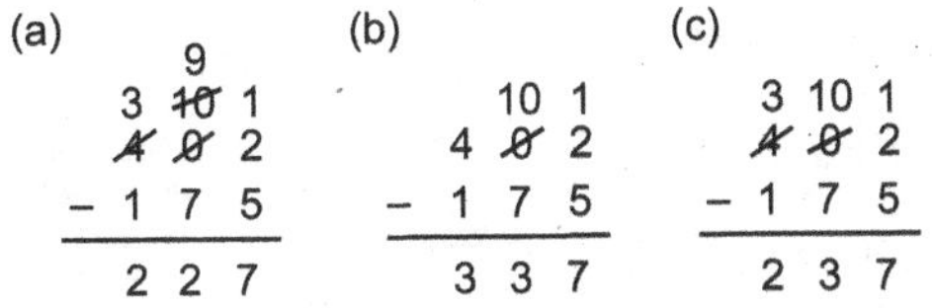

(a)
```
    9
  3 1̶0̶ 1
  4̶ 8̶ 2
- 1 7 5
─────────
  2 2 7
```

(b)
```
     10 1
   4 8̶ 2
 - 1 7 5
 ────────
   3 3 7
```

(c)
```
  3 10 1
  4̶ 8̶ 2
- 1 7 5
─────────
  2 3 7
```

Figure 2.13 Some of the anticipated student responses

Anticipated responses

    R1: (correct) (Figure 2.13a)
    R2: 402 − 175 = 373 (students subtract smaller from larger in each place)
    R3: Students don't decrement hundreds and tens when they regroup (Figure 2.13b)
    R4: Students correctly regroup hundreds to tens, but forget to decrement the tens when they regroup to the ones (Figure 2.13c)

Comparing and discussing

Post R4, R3, R1 on the board side-by-side.

Invite comparison and discussion: "What do you notice?"
"How could we use math to prove which response shows the algorithm being used correctly?"

Create a base-10 representation of the minuend (402) and use it to analyze the three solutions (see board plan).

When discussing the correct solution, focus on the important aspects of the algorithm: Because there are not enough ones in 402 to subtract 5, we need to regroup 10 ones from the tens place. There aren't any tens in the tens place, so we need to regroup 10 tens from the hundreds place, and then regroup from 10 ones from one of those.

Summing up

"Today we learned that when we have a zero in the tens place, we need to first regroup from the hundreds to the tens and then from the tens to the ones."

Board plan

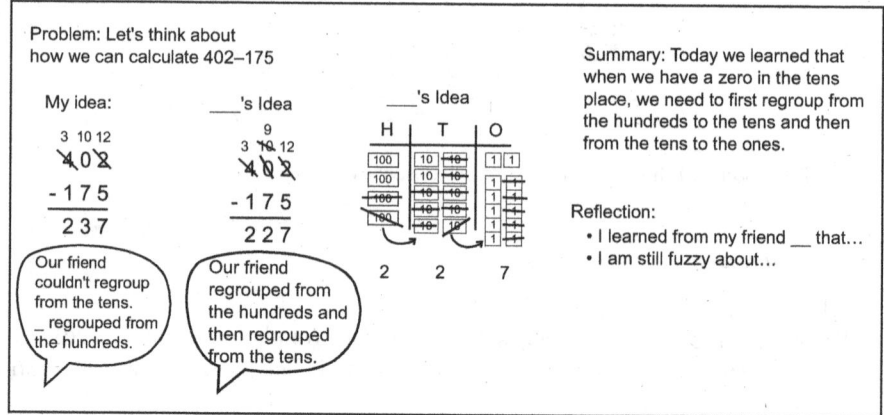

Figure 2.14 Board plan for the lesson

Observations from the lesson

Several things stood out for me about the design of this lesson. First, I was impressed that the team designed this as a problem-solving lesson, inviting students to figure out the process without being told, and I was impressed that the students seemed perfectly willing to engage with this problem. Their decision to open with a familiar task, 492 – 175, and then have students contrast it with the main task, 402 – 175, seemed to be effective at focusing students' attention on the challenging feature, the zero in the tens place.

I also noted that, rather than use base-10 blocks, they used a base-10 chart with rectangles labeled "100", "10", and "1" (see the center of their board plan, Figure 2.14). It struck me that this representation could actually be *less* abstract for some students than pictures of base-10 blocks, although they do share with base-10 blocks the limitations I described earlier.

As the lesson authors predicted, many students had difficulty with application of the algorithm in this problem, even though most of them used a base-10 chart. The most common errors were the ones identified as R3 and R4 in the plan – regrouping from 100s to 10s and then 10s to 1s but not decrementing the 10s in this second step and sometimes not decrementing the 100s either.

During the discussion portion of the lesson, as planned, the teacher invited two students to present their ideas, one incorrect, one correct. The incorrect one was R3 (answer 337). The teacher then used a comparison of the two solutions to try to help students understand the process of regrouping first from 100s to 10s, then 10s to 1s. I couldn't tell for sure how many students understood this, but the two students I was close to, who had made the same mistake (337), did seem to understand and corrected their work.

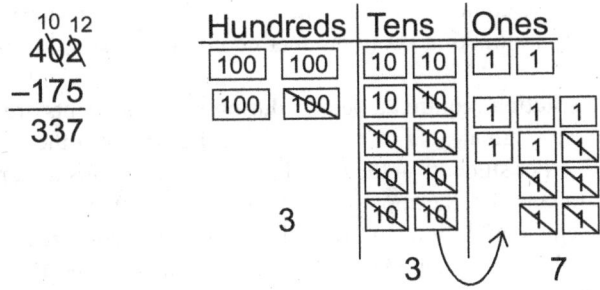

Figure 2.15 A student fails to decrement both the hundreds and tens when regrouping

What I learned

At this point, I felt that I had acquired several useful ideas about how to address this difficult topic with students. First, it was clearly worthwhile to devote an entire lesson to the problem of regrouping when there is a zero, as this is a conceptually difficult topic. I also saw how one single problem could be used to develop the idea, as opposed to using several examples. Third, I saw how prefacing the main task with a contrasting problem where there isn't a zero helped focus the students' attention on the new challenge of dealing with the zero. And fourth, I saw how asking students to compare incorrect and correct approaches pushed them to grapple with the meaning of each digit, its place value, and why regrouping was necessary.

On the other hand, I still wondered about the problem of not having some explicit representation of the subtrahend. When the problem was worked out using a place value chart, the rectangles (marked as either 100, 10, or 1) were crossed off in the same way, whether to indicate regrouping or subtraction. I wondered whether students who appeared to fail to decrement the hundreds (R3, Figure 2.13b) were in fact decrementing when they regrouped, but then getting confused when they went to subtract, thinking that because they had already crossed off one "100", they had already subtracted it.

But the mathematical insight I gained from this lesson came courtesy of the final comments provided by Dr. Tad Watanabe of Kennesaw State University. During the lesson, both students and the teacher had talked about 402 by saying, "There aren't any tens." That was indeed how I thought about it, and that was how the research proposal articulated it. But Dr. Watanabe pointed out that if one is able to view 402 flexibly, there are in fact 40 tens, and if you decompose one of them, that leaves 39 tens. If you think this way, then you have just one regrouping step instead of two.

This came like a lightning bolt to me. In my educational experience up to that point, I had seen numbers like 253 in only two ways: either as 253 ones or as 2 hundreds, 5 tens, and 3 ones. But this lesson was for me the beginning of realizing that any place could be used as the unit, depending on the need, and that greater flexibility in thinking about number could greatly simplify the algorithm for students.

The desired solution required a detour, during which students are supposed to put on hold their original goal to regroup from the tens into the ones place. As I had seen with other students in the past, I saw students in this lesson getting lost in that detour. In contrast, the approach suggested by Dr. Watanabe was strikingly elegant. If students could view 402 flexibly enough to recognize that there were in fact 40 tens, then they could stay focused on their goal and regroup one ten, leaving 39 of them.

Since this lesson, I have begun to recognize the value of being able to view numbers flexibly in many other contexts. For example, it is at the heart of the long division algorithm when we divide $246 \div 6$: since 2 (hundreds) is not divisible by 6, we move to the next place and divide 24 (tens) by 6. Similarly, $1.2 \div 3$ can be thought of as 12 (tenths) $\div 3 = 4$ (tenths), or $1.2 \div 0.4$ can be thought of as 12 (tenths) divided by 4 (tenths) $= 3$. It is also at the heart of scientific notation, where we think of a number as 2700 as 2.7 thousands, or 2.7×10^3.

Impacts on my own work

This research lesson has impacted my work with respect to this specific mathematics content, but also with respect to how I conduct *kyouzai kenkyuu* (research) when thinking about how to improve the teaching of a challenging topic.

In his final comments, Dr. Watanabe mentioned, as an aside, that Japanese textbooks explicitly devote time to teaching students to think flexibly about place value. I have now found one that does this; it asks students, for example, "How many tens are there in 230?" (Fujii & Majima, 2020). I now often consult this textbook series, as well as others, when I am thinking about how to improve the teaching of a topic, to broaden my perspective. So this experience has taught me to investigate other curricula when I am engaged in Lesson Study.

In terms of mathematics content, I encourage teachers of younger students, as soon as those students learn about hundreds, to devote some time to teaching students to express numbers in terms of different units, using problems like the one mentioned earlier. For a problem like $130 - 50$, whereas I used to see students cross out the 1 hundred and write a superscript 1 next to the 3, students now see this as $(13 - 5 = 8)$ (tens). When it comes to lessons involving calculations with decimals or with large numbers (e.g. hundred thousands), I encourage teachers to have students think of those numbers in different ways and use the perspective that makes the calculations easiest. I already mentioned how it helps with decimals; for large numbers – for example, to calculate $520,000 + 60,000$ – students learn that by thinking of these in terms of ten-thousands, they can easily calculate mentally.

I am extremely grateful to have had the opportunity to observe and learn from this lesson. Seeing the teaching ideas play out live with students made them much more vivid and concrete and easier to translate into my own practice, and seeing students struggle with doing two regroupings helped Dr. Watanabe's mathematical point hit home with me more than it would have if I had read it in an article or heard it in a workshop. This, I think, illustrates how Lesson Study can be powerful not only for those who participate in planning, but for observers as well.

Note

1 A full version of the lesson plan can be accessed at https://LSAlliance.org/Lessons

References

Common Core State Standards Initiative. (2010). *Common Core State Standards for Mathematics.* http://www.corestandards.org/Math/

Fujii, T., & Majima, H. (Eds.). (2020). *New mathematics for elementary school (2A).* Tokyo Shoseki.

Tondevold, C. (n.d.). *Stop using base 10 blocks to 'Teach' the algorithms!* www.therecoveringtraditionalist.com/stop-using-base-10-blocks-teach-algorithms/

Summary of Chapter II

What we learned about the contents we teach

Tad Watanabe

When teachers engage in Lesson Study, they can learn many different aspects of mathematics teaching and learning. However, one of the most crucial learning they can experience is learning, or deepening their understanding, of mathematics they are teaching. There is a consensus among mathematics educators that teachers need to understand mathematics they teach in a way that goes beyond the understanding that non-teachers may need. It is often referred to as mathematics knowledge for teaching (e.g., Ball, Thames, & Phelps, 2008). This mathematics knowledge for teaching does not necessarily involve more advanced mathematics, or mathematics professional mathematicians are researching. However, this understanding is critical to teach mathematics effectively, especially if teachers want to employ an ambitious pedagogy, like teaching through problem solving. For example, if you are going to teach multiplication in upper elementary grades, it is not sufficient that you can simply calculate products using the standard algorithm. Teachers need to understand different ways multiplication may be represented, such as area models and double number line representations, as well as affordances and limitations of various models. Teachers must also be able to solve problems involving multiplication in different ways and elaborate how different solution approaches are similar and different. Some of this understanding may be developed in preservice teacher education programs, but much of it must be learned throughout teachers' professional careers.

Different types of knowledge needed for effective mathematics teaching is represented in Figure 1.9 on page 24, often affectionately called the "egg diagram." Although the boundaries among these different types of knowledge are not always quite clear, this diagram illustrates the range of mathematics knowledge teachers need to be effective.

The five reports in this chapter illustrate that what teachers learn in Lesson Study might encompass multiple types of mathematics knowledge for teaching in the egg diagram. Brigid Brown and Jana Morse discuss challenges primary students face when they learn different types of addition and subtraction word problems. Brown's first grade students were dealing with take-from change unknown–type problems, while Morse's second grade students were working with compare problems in which the difference was described with "more," but the operation needed to find the answer was subtraction. In order to support their students overcome

DOI: 10.4324/9781003230915-11

these challenges, Brown's team explored the potential usefulness of storybook idea in helping students make sense of problem situations while Morse's team explored using different diagrams as students' reasoning tools. What teachers in Brown's and Morse's teams learned might encompass knowledge of content and students and knowledge of content and teaching. Thomas McDougal observed a third grade lesson on the subtraction algorithm where students must regroup across a 0 in the tens place of the minuend. He observed a different model used to represent 3-digit whole numbers in the research lesson, but he also learned that a more flexible way to see numbers might also be useful. Thus his learning may have encompassed specialized content knowledge and knowledge of content and teaching.

Both research lessons planned by Joshua Lerner's team and Aaron Bingea's team had two closely related objectives. In Lerner's lesson, they wanted students to understand that multiplication by a decimal number is an appropriate representation of a problem situation and think about ways to calculate when the multiplier is a decimal number. In Bingea's lesson, the team wanted students to understand why like terms in an algebraic expression may be combined by making use of contexts and justify it by using the distributive property. Although both lessons were engaging, they both felt they only met one of the two goals the teams set for the lessons. In Learner's lesson, students came up with the appropriate multiplication expression, but it was probably because there was too much scaffolding provided. In Bingea's lesson, students did not appear to feel the need to justify mathematically why like terms may be combined because the context of the problem made it obvious to them.

Through post-lesson discussions and own reflections, they both learned that the goals that might appear to be closely related might not be so closely related to their students for different reasons. The goals must be carefully sequenced with appropriate emphases. Their learning encompasses of knowledge of content and students as well as knowledge of content and curriculum.

As these five reports illustrate, participating in Lesson Study cycle allows teachers to deepen their understanding of mathematics they teach. Moreover, these reports also show that teachers' learning is not limited to what is discussed in the research lesson in the Lesson Study cycle. Each of the five authors discussed how their learning from the cycle impacted their work long-term. Of course, teacher learning in Lesson Study is not limited to learning of content. The following chapters will illustrate what other knowledge teachers may gain or deepen. Moreover, often what teachers learn leads to more questions to be investigated through additional Lesson Study cycles. The reports discussed in this and later chapters illustrate that, unlike typical professional learning workshops and lecturers, Lesson Study is a system of life-long teacher professional learning.

Reference

Ball, D. L., Thames, M. H., & Phelps, G. (2008). Content knowledge for teaching: What makes it special? *Journal of Teacher Education, 59*, 389–407.

III What we learned about lesson design and pedagogy

DOI: 10.4324/9781003230915-12

Designing lessons students lead

Numbers greater than 1,000, for grade 2 (7- and 8-year-old) students

Berenice Heinlein

I spent the first few years of my teaching career using the "I do, we do, you do" structure that I had been taught and following my curriculum with fidelity, but after a while I began to question if this pedagogy was right for my students. Although my students could restate math facts and scored relatively well on multiple-choice exams, it was clear that they were not developing math practices. When faced with new problems, my students waited silently for an explanation or became frustrated, demanding an explanation instead of persevering through problem solving. When asked to explain their thinking, my students sat quietly, waiting for me to give them the correct response. The way I taught didn't prepare students to reason through mathematics on their own, or to engage in a discussion about their learning.

I was starting to read about Teaching through Problem-solving (TTP), hoping for a solution. I had some understanding that TTP involved questioning, but I didn't know how to realistically use it in my class. I read about how students were supposed to solve a problem on their own and then learn about it through a discussion. I wondered how to design a lesson or unit that could make this pedagogy possible.

At the Chicago Lesson Study Conference in 2018, I observed a TTP lesson with a class of grade 2 students who were mainly English Language Learners. The content goal was for students to understand numbers greater than a thousand, usually taught in the following grade. The research lesson description stated:

> Students will receive a picture with 2,354 pennies. Students will grapple with how to count, name and write a number that is greater than a thousand. Students will recall that when they previously counted large numbers of objects in previous lessons, they made groups of tens and hundreds. Students will build off their knowledge from a previous lesson to see the efficiency in counting in groups, and even counting with groups of a thousand.

By participating as an observer in this Lesson Study, I was exposed to a lesson design where students were able to lead the lesson. I saw students confront a new problem and try to solve it independently and then discuss their ideas together by asking one another questions. I felt inspired by their perseverance and abilities.

DOI: 10.4324/9781003230915-13

This lesson completely changed my perspective on teaching and learning, showing me how students could develop Mathematical Processes such as persevering through problem solving and constructing viable arguments while developing new mathematical ideas as a classroom community.

Lesson plan (shortened[1])

> **Title of the lesson:** Let's Investigate Numbers Greater than 1000
> **Students' school:** Chavez Elementary School, Chicago, Illinois
> **Student ages:** 7–8 years
> **Instructor:** Rebecca Reddicliffe
> **Co-authors:** Ana Cabrera, Adriana Soto, Bethany Jorgensen
> **Date:** For the Chicago Lesson Study Conference on May 18, 2018
> **Goal of the lesson:** Students know how to name, count, and write numbers less than ten thousand in the base-10 value number system
>
> - Students learn the structure of a four-digit number and how to read these numbers (2,354)
> - Students recognize that ten groups of 100 make "one thousand" and two groups of 1,000 make "two thousand."
> - Students understand that putting "two thousand" and "three hundred fifty-four" together makes the number "two thousand three hundred fifty-four."
> - Students understand how to read the number that has two groups of a thousand, three hundreds, five tens, and four ones as "two thousand, three hundred fifty four."
>
> **Learning standard:** CCSS Math 2.NBT.A.1: Understand that the three digits of a three-digit number represent amounts of hundreds, tens, and ones; e.g., 706 equals 7 hundreds, 0 tens, and 6 ones.
>
> (Common Core State Standards Initiative, 2010)

Lesson flow

Introduction

Students are given a picture of many pennies with a familiar context. The school has a fundraiser called Pennies for Patients. This context provides a reason for students to count the pennies. The teacher explains the context and situation in which students will help another teacher by counting the pennies. Then, students estimate.

Posing the problem

"How many pennies are there? Think about the way you're counting the pennies."

Anticipated responses

    R1 (correct): Circle 100 pennies as a group and use the grouping strategy to count the total.

    – 23 groups of 100 and then 5 tens and 4 ones (23 hundred 54)

    R2 (correct): When counting up to 1,000, use a different color (or different circle) to show groups of a thousand.

    – 2 thousands, 3 hundreds, 5 tens, 4 ones (2,354)

Comparing and discussing

(Each letter shows a different anticipated student response to utilize in the discussion.)

A   I circled a group of 100 pennies and saw that there were 23 groups of 100 and then 54 more pennies, but I don't know how to say the number of coins.

    a   Place Value Chart (see Table 3.1)

    Table 3.1 An anticipated student solution using a place-value chart

Hundreds	Tens	Ones
23	5	4

    b   twenty-three hundred fifty-four

B   I circled a group of 100 pennies. When I made 10 groups, I made a group of 1,000 pennies and circled it with a big circle. I noticed there are two groups of 1,000, three groups of 100, five groups of 10, and 4 more pennies.

C   Two groups of 1,000 pennies make 2,000 pennies. Three groups of 100 make 300. And then we have 54 more pennies.

    a   How many groups of 100 pennies are there?
    b   How many groups of 1,000 pennies are there?
    c   Why do you think we made groups of a thousand?

Summing up

We learned that ten groups of 100 make "one thousand" and two groups of 1000 make "two thousand." We can use groups of 1000 to count more quickly. (Even though this is not our main goal, we feel that some students will view counting by using groups of 1000 as their main learning.) **OR** Main Goal: We learned that the number made of two 1000s, three 100s, five 10s, and four 1s is: Two thousand, three hundred fifty-four.

Board plan

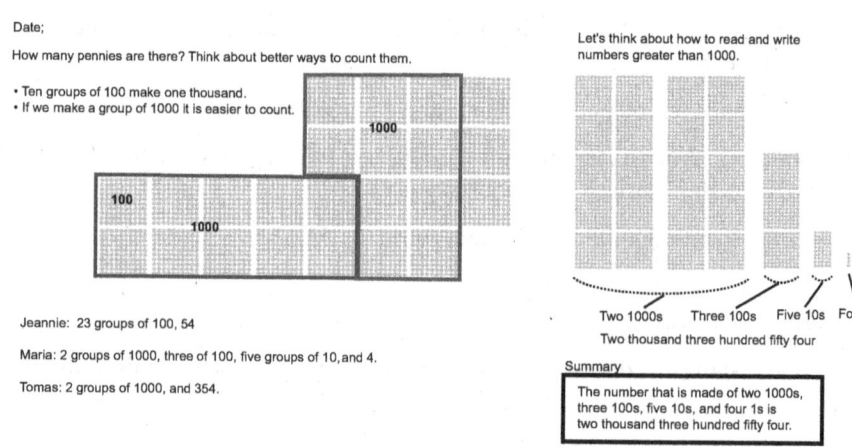

Date;

How many pennies are there? Think about better ways to count them.

• Ten groups of 100 make one thousand.
• If we make a group of 1000 it is easier to count.

Let's think about how to read and write numbers greater than 1000.

Jeannie: 23 groups of 100, 54

Maria: 2 groups of 1000, three of 100, five groups of 10, and 4.

Tomas: 2 groups of 1000, and 354.

Two 1000s   Three 100s   Five 10s   Four 1s

Two thousand three hundred fifty four

Summary

The number that is made of two 1000s, three 100s, five 10s, and four 1s is two thousand three hundred fifty four.

Figure 3.1 Board plan for the lesson

Observations from the lesson

The lesson began with the teacher helping students recall the strategies they used previously when counting large numbers of objects, mainly by making groups of tens and hundreds. Then, the teacher set the context for the problem in which it was important for students to count, and showed them the image of pennies. Students were excited by the challenge of counting such a large number. Students estimated the number of pennies and shared their reasoning. Then, students counted pennies independently, using highlighters and pencils to circle groups and to show their thinking. As a participant and observer, I noted that all students began working on the problem independently and quietly, showing their thinking by marking the paper directly. I focused on two students and saw that they both approached the problem differently, and that neither had a complete answer when it was time for comparing and discussing.

The discussion began with a "turn and talk" partner sharing where I heard two students close to me share how they approached the problem and what challenges they faced. Both were unsure if their answer was correct. To begin the whole-class discussion, the teacher selected a student to share his strategy. The student told the class what he did, step-by-step, as the teacher wrote it down on the board. The teacher prompted discussion by asking students to share questions or comments for the student who was sharing. Most students raised their hands, asking why he made specific decisions, asking him to count aloud to prove his idea, and sharing comments about how his work affected their thinking. After questioning, recounting, and prompting from others, the student determined that there were 23 hundreds and 54 more, but didn't know what number that made. The discussion seemed like a conversation between students, where they grappled with the idea. Eventually, students agreed that his reasoning made sense, but they still didn't know how to express the quantity of pennies.

The teacher then asked a second student to share, who identified two groups of 1,000. He counted 2 thousands, 3 hundreds, 5 tens, and 4 ones. The class followed a similar process for questioning, and students grappled with the idea that both could be true: You could say there are 23 hundreds, or 2 thousands and 3 hundreds, and both can be true. At some point, a student asked, "What do you think you would call this number?" Another replied, "I don't know, maybe something like 'two thousand and three hundred and fifty-four'?". After students discussed, the teacher closed with a summary writing that there were "2 thousands, 3 hundreds, 5 tens, and 4 ones" which make "two thousand three hundred fifty-four." Students wrote down the summary and wrote a reflection, sharing what they individually learned from the lesson.

In the post-lesson discussion, observers determined that students in this age group were able to access the content and were equipped to engage in problem solving even though this lesson addressed a standard for the next grade level. Some observers wondered if students fully understood why counting in groups of 1,000 can be beneficial. A major point of discussion was on how the teacher chose to present the total number of pennies at the end of the lesson. The "knowledgeable other" providing final comments agreed with the teacher's choice not to write it as a four-digit number. He emphasized that the lesson engaged students and provided the space for them to grapple with new ideas and to talk about mathematics. By showing patience and not giving students the answers, the teacher created the space for them to think about and discuss mathematics in the future.

Figure 3.2 Students share ideas about counting by 100s and counting by 1,000s

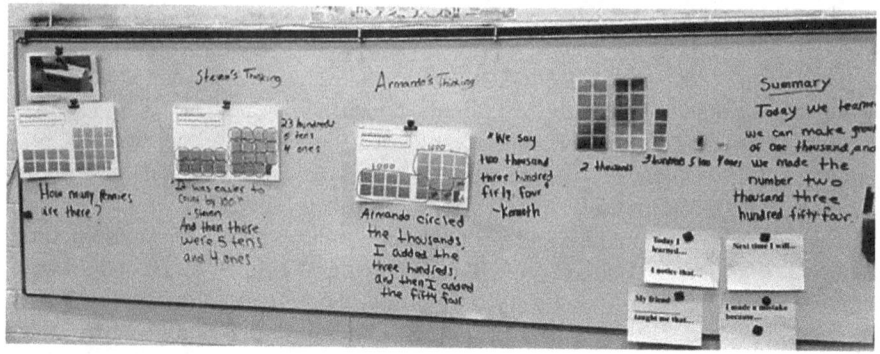

Figure 3.3 The board at the end of the lesson

What I learned

I was amazed by the way these students tackled the problem, challenging one another's ideas in the discussion. I also knew that my students could do the same if given the chance. They would be able to persevere through problem solving and create viable arguments if given an opportunity to develop those skills. I observed that TTP lessons empowered students to lead their learning. I felt inspired and equipped to make changes to my teaching.

At the time, my standard practice began with showing a decontextualized equation, or a word problem that the class would dissect together. When the research lesson teacher posed the problem, she used a context and method that all students could access. The problem seemed straightforward as students had experience with counting pennies, and the challenge exclusively focused on the topic of new learning. Although the teacher did not show students how to count numbers above 1,000, they were expected to think and reason through possible solutions. I was delighted to see a class of 7-year-old students excited to start the problem.

The teacher in the research lesson didn't walk around to help push individual students in the right direction. When students got stuck, many continued to reason through the problem, counting again and checking their work. My own approach at the time was just the opposite. I was telling students about mathematical ideas that they hadn't experienced and therefore could not appreciate or fully understand. By jumping in to help my students during independent work, I was taking away opportunities for them to deeply think about math while unintentionally communicating to them that struggle is an indicator that something is wrong. I learned that it was important to designate time for students to engage in productive struggle.

My classroom culture at the time prioritized answer-getting, and there was an emphasis on "the correct answer." As a result, my students were reluctant to make mistakes, take risks, and grapple with new content. When the students in the research lesson shared how they approached the problem with a partner, they didn't seem upset or embarrassed about not having an answer. Clearly, it was acceptable, even normal, not to know the answer by the end of the independent work time. The discussion began before even a single student had a completely correct response, and students didn't seem to mind. In the whole-class discussion, students felt comfortable asking one another questions. I could see how an emphasis on thinking through mathematics and learning together, rather than on answer getting, cultivated a classroom culture that was conducive to student-led learning.

In my standard practice at the time, I often tried helping students by coaching them through any struggle. In the discussion portion of the research lesson, I noticed there were many moments where I would have interjected had I been the teacher. Selected students shared strategies and other students in the class facilitated the discussion through questions and comments. One student initially counted incorrectly, and reconsidered their answer during the discussion. At times, the class discussion paused while students pondered a new idea or grappled with a question. Instead of taking over the lesson and giving students an answer, the teacher let them think through challenges, further reinforcing that it is important to problem solve and that it is all right not to know an answer. The teacher chose to allow the class to reason through the problem together, instead of sharing the right answer or teaching directly. Those moments often led to students learning new and important ideas about mathematics. I learned that it is important to create lessons where students engage with mathematics and create meaning together.

By the end of the lesson, students were able to count the pennies by counting groups, including groups of thousands, and it was evident that they learned something new. Together, they determined that 23 hundreds was the same amount as 2 thousands and 3 hundreds, a concept that my 9-year-old (grade 4) students often struggled to understand. By engaging in productive struggle together, students created meaning, solidifying foundational place value concepts. I was impressed to see a class of second graders lead the lesson through their discussion and learn mathematics by reasoning together.

Impacts on my own teaching

Observing this research lesson led me to completely change my approach to teaching mathematics. It became the foundation for my pedagogical beliefs on mathematics instruction. I contacted co-workers to share what I learned and started participating in Lesson Study with others in my school to learn more. I began designing lessons that gave students the opportunity to persevere in problem solving by adopting the TTP lesson structure that I observed in this lesson. I focused on encouraging students to reason through new ideas instead of prioritizing answer-getting. Although making these changes to my instruction felt strange, I saw an impact soon after implementation. I now use TTP for math instruction, and my data shows increased student learning as a result. I learned so much by observing a single lesson, so I now attend research lessons as often as I can.

Note

1 A full version of the lesson plan can be accessed at https://LSAlliance.org/Lessons

Reference

Common Core State Standards Initiative. (2010). *Common Core State Standards for Mathematics*. www.corestandards.org/Math/

Creating meaning with data and graphing

Collecting, graphing, and interpreting data, for grade 2 (7- and 8-year-old) students

Meghan Smith

On a cold Chicago evening in 2017, where the sun set at 4:30 PM, a team of dedicated teachers met to begin planning for a research lesson that would eventually be taught in front of other educators at the Chicago Lesson Study Conference. The team consisted of two second grade teachers, a first grade teacher, and a special education teacher. We began our Lesson Study process by discussing topics that were historically difficult to teach or difficult for students to understand. Topics such as missing addend and adding multi-digit numbers came up, but what resonated most with this team of teachers was how to teach topics that seem to always be presented as direct instruction. Data collection and graphing was just such a topic. So we decided to select this as the topic of our research lesson.

In our research and analysis of different approaches to teaching data collection and graphing, we found that most curricula taught this topic in a very direct way. This was even true of curricula that are usually problem-solving- or inquiry-based. All of the texts we looked at called for the teacher to model how to take data/information that is scattered and place it into a bar graph or pictograph.

We wanted students to see the usefulness of a visual representation rather than just graphing a data set because the teacher told them to. We thought it would be important for students to have a connection to the data in order to increase their interest in the task and help them see how math could be useful in their life. And, we wanted, as much as possible, for the important ideas about graphing to come from the students.

With these intentions in mind, we made a strategic choice for the data set we would use in our lesson. At the end of each day, students are allowed to have free choice time. There are six categories for free choice time that students could pick from: iPads, Magna Tiles, Computers, Play-Doh, Coloring, and Math Tools. Students write their name and their free choice selection on a sticky (Post-it®) note. We thought we could use these free choice selections as the basis for a lesson on graphing. Because the data was about the students themselves, they would feel connected to it. Although the Common Core State Standards for grade 2 only call for four categories of data, we thought students would better see a purpose in representing data with a bar graph if there were more than four categories. And the sticky notes would lend themselves to being lined up to make a bar graph. We just needed enough data to make it hard to analyze without the organization of a

DOI: 10.4324/9781003230915-14

graph. There are 28 students in the class, but there are usually fewer than 28 data points each day due to student absences and other factors. So we decided to use five days' worth of student selections.

I had two key learnings because of this research lesson. My biggest learning was that students were able to come up with the idea of a bar graph on their own, without the direct instruction of the teacher. Along with this, my biggest takeaway about math instruction in general is even when it first seems as though a topic has to be taught through direct instruction, I have learned that if I trust my students, there is usually a way to have the ideas come from them.

Lesson plan

Title of the lesson: How Can We Organize This Information?
Students' school: O'Keeffe School of Excellence, Chicago, Illinois
Student ages: 8–9 years
Instructor: Meghan Smith
Co-authors: Caileen Brett, Alyssa Frollo, Emily Kempe
Date: May 12, 2017
Goal of the lesson: Students will understand how to use a visual model to represent data. Students will understand the purpose of displaying data on a bar graph. Students will understand that bar graphs help find information quickly. Students will understand why a bar graph is useful.
Learning standard: CCSS.MATH.CONTENT.2.MD.D.10: Draw a picture graph and a bar graph (with single-unit scale) to represent a data set with up to four categories. Solve simple put-together, take-apart, and compare problems using information presented in a bar graph.

(Common Core State Standards Initiative, 2010)

Lesson flow

Introduction

Teacher will display Post-it® note data on the board, scattered randomly, and students will explain what the notes represent and how they collected this information.

Invite students to generate questions one might ask about the data, e.g. "Which activity did students choose the most?" or "Did more students choose Play-Doh or Magna Tiles?"
"Why is it so hard to answer these questions?" (The notes are all over the board, we can't read them, it is hard to see.)

Posing the problem

"How can we organize the sticky notes to make it easy to answer the questions we
   have about our choices?"

Teacher will write the task/guiding question on the board.

Students will think individually, then work on the problem in groups. Students will
have a "mini" set of post-it notes at their desk that reflects the same data on the board.

Assessment: Do students understand the task? Are students eager to solve the
problem?

Anticipated responses

1   Students create a written list with the data.
2   Students make a tally chart based on the data.
3   Students group same data together.
4   Students group the same data together and categorize in a graph type of display.

Assessment: Are students able to tackle the problem? Do students have a solution
method that they are ready to share at the discussion? What are the different ways
groups are thinking they should organize the data? Do the sticky notes help students
gain insight into the problem? Are students using accountable talk in groups?

Comparing and discussing

For this lesson, the board will serve as a "think pad" for students to collaboratively work on moving from random data to creating a visual display that helps us
understand the data.

Teacher Support: Ideas to focus on during the discussion

- We want the progression to move from randomly organized data, to grouped
  data, to a bar graph of the data.
- We plan to use the sticky notes to re-order the data in an organized way.
- Main point about the sticky notes is that each note represents an individual
  data point, but the data points together create a bar.
- Labeling the graph to explain what each bar and axis represents.

Why is it hard to answer questions about this information with just the sticky
   notes?

Why is it easier to answer questions about the data when it is grouped together
   (we can easily count how many)?

Is there an even better way we can organize the data than just grouped together by
   the same categories (grouped in bars like a graph)?

How can we show what each bar represents (labels)?

What does this graph help us do? (Compare amounts, see how many total, see
   how many for each category.)

Assessment: Are students defending their ideas? Are they responding to each other's ideas? Are students able to explain why it is hard to analyze random data? Are students able to justify why it is easier to understand the information when it is categorized? Are students able to answer questions about the data when it is grouped together? Are students able to label the graph and explain the different components/what they mean? Are students able to express why the graph is helpful?

Summing up

New learning statement: As a hardworking math class, today we learned that we can create a bar graph to help us organize and understand data.

Students will write a reflection in their journals about their experience with the lesson/task today.

Assessment: Is the new learning statement reflective of what students learned today? Are students able to express what they learned/who they learned from?

Board plan

1   Begin with the notes scattered randomly.
2   Group the notes by activity.
3   Arrange the notes in vertical columns, with labels and a numerical scale as the vertical axis.

Observations from the lesson

Some key things that observers noticed was that student engagement was high at the beginning of the lesson and students were very motivated and curious about the data. Observers also noticed that many students were categorizing and grouping the data. There was one group that started to line up the Post-its, but was constrained by the size of the blue paper (the blue paper was an 11x17 piece of construction paper to be used as a work mat). There was another group who lined up the Post-its, but then stacked them once they had a line.

Observers noted that students were able to discuss with one another and worked well together in groups. Some students had trouble articulating their thoughts, and many students might have been confused with what it means to organize as many of them only categorized. During the discussion, many people were wondering about the context and expressed that it might have been more motivating for students if they had a purpose for creating the display.

Key points from the final comments

Dr. Akihiko Takahashi from DePaul University provided the final comments for our lesson. His comments pushed our thinking in what we think was a helpful direction. He remarked on how when the teacher trusts their students and believes in their students, lots of learning can happen. He said that the first thing he

Figure 3.4 Students line up the sticky notes in columns

wondered when looking at the Post-it® notes was how many there were and how this is a very natural question for students. His feedback was that this lesson was a very full lesson that could have been broken up over two days. The first day could be the data collection and categorizing in groups and the second day could be organizing the categories into a bar graph. He also suggested changing the context to create a problem for the students that would be solved by making a data display. For example, I asked questions of the students during the discussion where a bar graph would be helpful for them to interpret and discuss the data. Dr. Takahashi suggested this question frame the lesson as opposed to being just a point of discussion during the lesson. He said that the main point for creating a visual display of data would be so someone else could easily understand it. Without this context, it is hard for students to move from stage 2 (categorizing) to stage 3 (graphing). For example, with the bar graph lesson, the students came up with the learning, and so moving forward it is more likely that they will gain an appreciation for a bar graph versus just graphing because the teacher tells them to.

What I learned

This research lesson was taught at the Chicago Lesson Study Conference as a public lesson. I was hopeful that conference attendees would see the true value of teaching in this way. At the same time, it was very difficult for me as the teacher to take a step back and truly allow the students to be the center of the learning. There were many points during the student discussion, which went on at great length, where I had to remind myself to allow time for the students to make sense of the data and construct the bar graph organically. When I let the students have the time and space to really uncover the learning on their own, their learning from this lesson and unit was much deeper than in previous years.

Because of Lesson Study and researching this topic at a deep level, I learned so much about data and graphing than I could have anticipated. When we first started our research, I was not sure how we could make this more of an inquiry-based topic and was worried that this was a topic that might have to be taught in a more direct way. Through research, I learned how to make data meaningful to students especially for something abstract like a bar graph.

I also learned the importance of the board plan and have never spent more time planning a specific part of a lesson. In *Teaching Mathematics through Problem Solving*, Takahashi (2021) discusses the four types of neriage. Our plan is Type 1, "develop a new idea by showing the progression of thought." But whereas the usual progression for this type of board work would be showing Idea 1, Idea 2, Idea 3, etc. alongside each other, our lesson used the board as a collective "think pad." We tried to apply the research about concrete, pictorial, and abstract in how the data would be sorted by the kids on the board, eventually depicting a bar graph that would evolve alongside the class discussion.

Impacts on my own teaching

This lesson impacted the mindset I bring to each topic and lesson in the curriculum. Even when I encounter a lesson that seems to require direct instruction, I look for ways to create the conditions that allow students to develop the ideas themselves through problem-solving. I have come to believe that almost any topic can be approached by Teaching through Problem-solving, in which students are deeply and authentically involved in creating the learning. Even for the few topics where this is not possible, things like notation or other conventions, I still try to first allow students to appreciate the need for or value of this new learning.

This lesson also allowed me to see the importance of anticipating student responses and the importance of the board work, which completely changed how I plan my math lessons. With regards to anticipating the responses, I was reminded of the importance of completing the task myself in the exact way it will be presented to my students. I did not anticipate students running out of space on the blue paper because when we completed the task together as a team we did not use the blue paper. As small of a detail as we thought this was, it ended up impacting some of our students and how they interacted with the task. In my teaching now, I often find myself thinking about the anticipated student responses first and how I will sequence them on the board when I plan my lessons and units. This helps me to think about the learning through the eyes of the students, which in turn helps me meet the students where they are.

References

Common Core State Standards Initiative. (2010). *Common Core State Standards for Mathematics*. www.corestandards.org/Math/

Takahashi, A. (2021). *Teaching mathematics through problem-solving: A pedagogical approach from Japan*. Routledge.

Boardwork as a roadmap

Introducing subtraction, for kindergarten (5- and 6-year-old) students

Aubrey E. Perlee

One of the most crucial things that I have learned through my Lesson Study experience is the significance and importance of boardwork. Thinking about and planning boardwork has transformed the way that I think about teaching and plan for student learning. In this section, I will share a research lesson that I taught at the beginning of a subtraction unit and how planning for the boardwork impacted the lesson and student learning.

The kindergarten team developed this lesson based on two patterns that we noticed in previous years. First, it was challenging for kindergarten students to fully understand what the subtraction symbol means. It was typical that one or two students may know the symbol ahead of time and write it in their number sentence. However, we had not found a student who understood the subtraction symbol and could explain *why* they used it. Second, the subtraction unit falls immediately after our addition unit, and students have often struggled to make the mental shift to using a symbol that was different from the addition symbol, which they became very comfortable and confident using.

The goal of the research lesson was that students would begin to understand why and when the subtraction symbol is used and, by the conclusion of the lesson, would be able to represent a takeaway situation with a number sentence.

Design of the unit

This unit introduced the students to subtraction. During the first lesson of the unit, students were presented with stories involving subtraction situations. All these stories were physically acted out to provide a clear context for what was happening (ex: There are four students doing jumping jacks. 1 sits down. How many students are left doing jumping jacks?). Students were tasked with orally retelling and acting out the story and then manipulating their counting blocks on the board to represent the stories.

DOI: 10.4324/9781003230915-15

The research lesson was the second day of the unit. The goal was for students to build upon their learning from the day prior by attempting to write a number sentence that represented their block manipulation and thereby the story. The subtraction symbol was introduced for the first time during this lesson.

Students had experiences representing addition stories with blocks and with number sentences. They knew that they were accountable for explaining how each component of a number sentence relates back to the original story. In this lesson, students were given the task of composing a number sentence that tells a new kind of story: a subtraction story. This lesson got students thinking about numbers and symbols they needed to represent the problem, as well as the order of those numbers and symbols. After having time to think about and attempt to write a number sentence, followed by share-outs arranged strategically to arrive at the correct order of numbers (5 2 3), the teacher introduced the subtraction symbol at a time when students were, essentially, asking for a symbol to represent a takeaway situation.

Lesson plan

Title of the lesson: What Is Left and What Is the Difference?
Students' school: Dr. Jorge Prieto Math & Science Academy, Chicago, Illinois
Student ages: 5–6 years
Instructor: Aubrey E. Perlee
Co-authors: Sari Freier, Katie Arnold, Sarah Bentley, Maria Estrada, Jennifer Sparacino, Andrew Friesema
Date: March 14, 2018
Goal of the lesson: Students are able to represent a "take away" situation with a number sentence.
Learning standards:

- CCSS Math K.OA.1: Represent addition and subtraction with objects, fingers, mental images, drawings, sounds (e.g., claps), acting out situations, verbal explanations, expressions, or equations (Common Core State Standards Initiative, 2010).
- CCSS Math K.OA.2: Solve addition and subtraction word problems, and add and subtract within 10, e.g., by using objects or drawings to represent the problem (Common Core State Standards Initiative, 2010).

Lesson flow

Introduction & posing the task

Display the picture in Figure 3.5.

Figure 3.5 Picture showing the subtraction situation

"Who can think of a story to tell about this picture?"

Students will share the stories they came up with. Class agrees upon a single story. Students glue the picture into their notebook.

"Now tell the story to yourself using your counting blocks." Students sit at seats and tell story to themselves as they manipulate the blocks.

"Turn and tell the story to a neighbor using your blocks."

"Your task today is to write a number sentence that matches our story."

Teacher Support: If students do not recognize that the girl is taking the fish out (and likely think she's adding them to the tank), draw attention to the movement lines, as well as the bowl she's putting them in.

Points of Evaluation: Are students able to effectively communicate the story to their friend and model the situation with their blocks?

Anticipated student responses

a. tells the story only with blocks
b. 5
c. 3 + 2 = 5
d. 5 2 3
e. 5 + 2 = 7
f. 5 + 2 = 3
g. 5 (blank) 2 = 3
h. 5 2 3
i. 5 − 2 = 3 [correct]

Comparing & discussing

For each response shared, students will come to the board, model the story with the large counting blocks and write their number sentences on the board. The class will discuss each number sentence as it's shared and begin to compare/contrast the number sentences after each is presented.

R1: $3 + 2 = 5$

Teacher Support: Draw students' attention to the numbers and how they're related to the story. "Let's retell the story using this number sentence." Help students to realize that this number sentence does not tell the story we agreed on at the beginning of class. The story written is about adding and our story was about taking away.

R2: $5 + 2 = 3$

Teacher Support: Draw students' attention to the numbers as they were recorded and how the number sentence does not tell the story: "Does $5 + 2 = 3$? Are there 7 fish somewhere in the story? What does the + symbol mean?"

R3: 5 (blank) $2 = 3$

Teacher Support: Have the student retell their story using their number sentence. "Does 5 take away $2 = 3$? Does 5 take away $2 = 3$ tell our story?" Now our number sentence matches our story, but we're missing an important piece, the subtraction sign. Introduce the subtraction symbol.

R4: $5 - 2 = 3$

Teacher Support: Our thinking is the student who uses the subtraction symbol at this point might not have a sophisticated understanding of its meaning. They may use it correctly, but they do not fully understand why. The symbol for subtraction has not yet been taught, but it is anticipated that some students may have encountered the symbol outside of class.

Point of Evaluation: How is the comparison and discussion benefiting students? Are students actively listening to their peers throughout the discussion? Are students able to explain each other's ideas? Is there evidence that they are developing their mathematical practices (justifying their conclusions & communicating them to others, responding to the arguments of others)?

Summing up

"We know that we can use a subtraction symbol for stories where we need to take away a group of objects."

On the right side of their notebook, students write the minus symbol in their notebook and the number sentence that matches our story.

Board plan

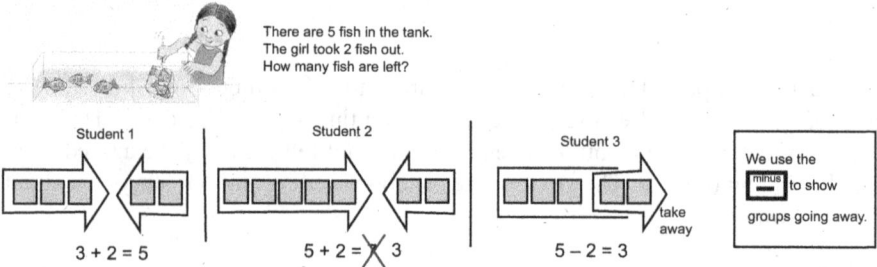

Figure 3.6 Board plan for the lesson

Observations from the lesson

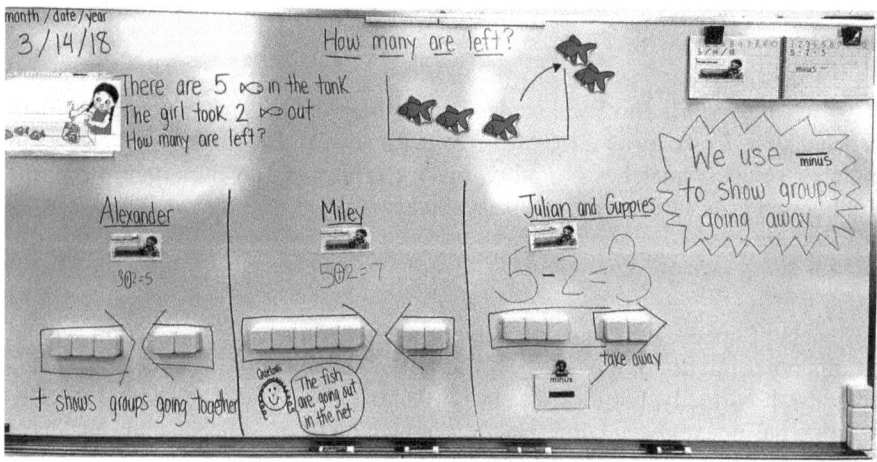

Figure 3.7 The actual boardwork from the lesson

At the beginning of the lesson, the students weren't seeing the picture as representing a "take away" situation. Many students saw a group of 3 fish in the tank and the girl putting 2 more fish in the tank. This is a logical conclusion, particularly since we just finished a unit on addition. We needed to get to a place where everyone agreed on the story, so I made the decision to draw the tank on the board and put actual fish pictures in it that I could manipulate and move to physically show the students what was happening. I could have acted this out but specifically chose to memorialize this on the board so students could return to it throughout the lesson to use it as a reference if they were confused about the story. This helped to solidify their understanding.

As students were telling the story back at their tables, virtually everyone was able to physically manipulate their counting blocks to start with 5, push 2 away and then touch the 3 counting blocks that were left. Observers noted that one student looked back at the picture I drew on the board and reenacted the story to show his 2 counting blocks jumping out of the tank. This was a key step toward understanding that this type of story was different from what they learned about in addition.

Observers noticed that the most common error was Miley's ($5+2=7$). The students knew that they needed to use symbols in their number sentences and the addition symbol is the only one that they knew represented the movement of two groups. When Miley told the story to herself at her table with her blocks, she knew that there were 3 fish left when she would touch her 3 remaining counting blocks. However, once she was writing her number sentence, she focused on her addition fluency (automatically knowing that the sum of $5+2$) to solve the problem. She either did not realize or chose not to acknowledge that this answer of 7 was different from the 3 fish that were left when she initially told the story.

Julian wrote (5 (blank) 2 = 3) in his notebook. He understood that the 2 fish were "going away" from the 5 fish, but he said, "I don't know what symbol that is", which was the perfect segue into introducing the subtraction symbol. When I showed students, many of them exclaimed that they had seen that before, proving that they had some knowledge of it but did not fully understand what it meant.

This lesson was developed so that students would progress from the concrete representation (the fish picture) to the semi-concrete (the counting blocks that represented the fish) to the abstract (the number sentence). During the comparison and discussion, I focused on asking the students questions where they had to re-contextualize the abstract number sentence back to the counting blocks and the fish picture. There was significant evidence that during this time, students were doing this, even without my prompting, as they were considering their friend's number sentence and describing why they agreed/disagreed with it. One student, Quetzali, said, "The fish are going out in the net" when she was describing why Miley's addition symbol didn't make sense with this story.

Ultimately, the team found that the lesson was largely successful with respect to our mathematical goals for students. From planning and teaching this lesson, I strengthened my understanding about how a good board plan can be a valuable teaching guide for me, leads to the creation of a valuable reference for my students and is a representation of the mathematics created by our class that day.

What I have learned

Boardwork as a roadmap: for the students

Through Lesson Study, I have become attuned to the relationship between myself, student responses, and the boardwork. Before I was introduced to Lesson Study, I had never even considered anticipating student responses in such detail, and I certainly had not considered the relationship between the boardwork and student learning. My experiences with Lesson Study have taught me to view my role as a facilitator, and, therefore, I think strategically about what should be written on the board. What I choose to write on the board determines the flow of the lesson, and effectively determines what the students will learn that day.

As the facilitator, I focus on helping my students have active, meaningful interactions with our boardwork. The board is not a sacred place where only the teacher can record thoughts or manipulate objects. The boardwork is the visual representation of the students' thinking and learning for that specific lesson, so it is important that the students feel a sense of ownership. This begins when I choose a student to come up to the board. That child models with blocks, takes the marker, writes their number sentence, and fields questions from friends about their strategy. Then, they can call on friends to again model what they did with their counting blocks or to explain their number sentence. This helps engage all students because kindergarteners love to share their thinking and have the full attention of everyone in the room.

Since kindergarten students are just learning how to read, most of the board-work consists of pictures, manipulatives, and arrows to show movement. I write some words on the board, but I am selective in what I choose to record because it is essential that students be able to read and internalize it. I prioritize recording direct quotes from the students (such as Quetzali's analysis of Miley's number sentence) and key differences between each number sentence that was shared.

I also learned that it is sometimes useful to leave the boardwork up until the next day. If a student was absent, their friends can explain what we learned in math the previous day. This is helpful for the absent student, who can see a physical representation of what occurred, as opposed to just an auditory recall. Additionally, the students who explain what happened internalize the new learning and will hopefully transfer their knowledge to the upcoming lesson.

Boardwork as a roadmap: for the teacher

In my first research lessons, I created a board plan almost as an afterthought to designing the lesson. But as my experiences with Lesson Study continued, I came to appreciate how thinking about the board plan early on was a useful tool for planning the lesson in the first place.

As a result, I used the boardwork as a roadmap when designing this lesson with my team (and generally, when I plan all lessons). We began with what the end goal of the lesson was or the new learning for the day and mapped out what we hoped the final share out would be. Then, we thought about the likely anticipated responses and determined what the ideal boardwork would be. We continued to backwards design the lesson from there.

As I plan, I keep in mind what the boardwork will look like and think about if the progression of sharing various strategies makes sense. As students are independently solving their problem in their notebook, I often have a picture of my ideal boardwork close by so I can record which strategy or number sentence students are writing. I keep this with me during the comparison and discussion so I can reference it to help me stay on track.

The boardwork is a well-laid plan, but of course it should be flexible so that it is representative of the mathematics the students are doing. The longer I teach through problem solving and the more I practice anticipating student responses, the more closely aligned the actual boardwork looks to the boardwork I planned, as with this research lesson.

Boardwork as a roadmap: as a reference

After I realized how important boardwork was for student learning and teaching planning, I felt that the next logical step was to begin taking photos of the boardwork each day. I compile these into a document at the end of each unit. These documents have been a great way to be able to look back at each of the teaching points and mathematical learning that happened throughout the unit.

My kindergarten teammates and I use these as a reference point when planning in subsequent years. We typically follow the same scope and sequence, and reference our previous lesson plans; however, I have found it beneficial to have a visual representation of the actual learning that occurred for students and not just the plan itself. Our curriculum has some boardwork examples mocked up digitally, but I have found it more powerful to keep a record of the actual learning that happened in our classroom; with my students' thoughts, misconceptions, and growth recorded. We can see how the flow of each lesson went in previous years and what errors commonly occurred, which helps us make decisions about adjustments we may need as we are planning for our current class of students.

I also print the boardwork pictures and typically review them each morning for the lesson I'm about to teach. This is a good, quick refresher and helps me think through anticipated responses, and the order they should be shared out on the board.

Lesson Study helped me to think strategically about how I use the board during a lesson. This has led to a positive impact on student learning and shifted the way I think about and plan lessons. Boardwork is now an anchor and reference point throughout each of my lessons and has become an integral component to teaching and learning in my classroom.

Reference

Common Core State Standards Initiative. (2010). *Common Core State Standards for Mathematics*. www.corestandards.org/Math/

Summary of Chapter III

What we learned about lesson design and pedagogy

Thomas McDougal

When people first learn about Lesson Study, they often misconceive its purpose as being about developing a "perfect" lesson. With more experience, however, one understands that this is incorrect for two reasons. First, there is no such thing as a "perfect" lesson. Children are different, and so a lesson that works well for one group of students may not work for another group, nor would it work well for the original students if used at a different point in their learning. Second, and more important, the purpose of Lesson Study is to improve teaching and learning generally; the research lesson should be an exploration of new pedagogical ideas and practices. Whether the lesson "works" for a particular group of students or not, we expect to learn something new about how future lessons should be designed and implemented.

This chapter gives three examples of how Lesson Study can lead to dramatic changes in how teachers teach mathematics.

Berenice Heinlein (Chapter 3.1) recognized the inadequacy of the "I do, we do, you do" approach that she was using, but was unsure about how to change. She observed a grade 2 research lesson at a Chicago Lesson Study Conference that employed teaching through problem solving (TTP), and was impressed by students' perseverance, independence, and conversation. That experience helped her understand what teaching through problem solving was and inspired her to begin trying to teach her own students through problem solving.

Meghan Smith (Chapter 3.2) had been using teaching through problem solving in some of her lessons already, but had doubts about whether graphing could be taught to grade 2 students that way, since the various curricula she and her team investigated all took a direct instruction approach. Nonetheless, they decided to try to create a lesson in which they would let students think about how to display some data in a way that made it easy to analyze. Their success with the lesson left Smith dedicated to seeking ways to use TTP to introduce all new math concepts.

Finally, Aubrey E. Perlee (Chapter 3.3) learned through Lesson Study about developing a board plan as part of crafting a research lesson. Eventually, the board became a central organizing focus of her work. Planning how she will use the board is an integral part of daily lesson planning. She uses the board strategically during her lessons to help students understand their task and to capture the progression of their ideas. She sometimes uses the previous lesson's board work to

DOI: 10.4324/9781003230915-16

help her students review their learning (and to help any students who were absent to catch up). And she takes pictures of the board each day to document the flow of learning during the year and to aid her planning in the following year.

Each section contains a lesson plan that is in itself an important contribution to the math education community. In Chapter 3.1, the lesson plan describes how one could introduce grade 2 students to the idea of thousands as a new countable unit (like hundreds and tens). In Chapter 3.2, the lesson plan presents a way to lead grade 2 students to see how they could arrange Post-Its to make bar graphs that make it easy to answer questions about the data. In Chapter 3.3, the lesson plan shows how a new type of situation – take-away – can lead kindergarten students to experience the need for a new mathematical symbol – the minus sign – and to be excited to learn about it.

But the purpose of Lesson Study is not to create a perfect lesson – no such thing exists, since each group of students is different – or even necessarily a *good* lesson, as much as we may try. The purpose of Lesson Study is to drive teacher learning. Although we may marvel at the pedagogical beauty of the lesson plans, the teachers who contributed to this chapter make a powerful statement about how Lesson Study has impacted their work with other lessons, with all their students. It is these broader impacts of Lesson Study that deserve to be celebrated.

IV What we learned about student

DOI: 10.4324/9781003230915-17

An upper elementary teacher learns about making ten in first grade

Making a new ten, first grade (6- and 7-year-old) students

Kari Laux

It's October in my fifth grade classroom, and we are in the middle of a unit about decimals and place value. Students have been asked to solve $4{,}268 \div 10$. Some students start to draw out place value charts. Some begin repeated subtraction. A few try long division. Others quickly jot down 426.8 and wait. This is the power of place value. For those who understand the base-ten number system, this question takes seconds. For those who haven't internalized the idea that when dividing by ten, each digit moves over once place value to the right, this problem could take a whole class period.

Having primarily taught in upper elementary grades (fourth and fifth), I have long seen the repercussions of a limited understanding of place value. The base-ten number system is complex, and for students who possess a limited understanding, many parts of upper elementary math learning (and beyond!) can be challenging. Without a deep understanding of place value, students are not comfortable with and do not fully trust the numbers enough to manipulate them. Not trusting the base-ten system leaves student mathematicians wary of algorithms for addition, subtraction, and even multiplication and division. Or perhaps worse, it leaves them applying these algorithms blindly from rote memorization. Without deep knowledge of base ten, students lack number sense and the ability to check the reasonability of their answer or estimate. Mental math is challenging. Decimals are incomprehensible.

For students who can think flexibly about base ten and feel comfortable with it, there are many benefits. They are flexible problem solvers. They feel confident extending their understanding of numbers to new, never-before-seen situations. For example, having never tried four-digit addition, they might feel comfortable doing so based on what they know about two-digit addition. They use friendly numbers to make mental calculations and to check the reasonability of their answers. They might catch an error in their multi-digit multiplication because they can estimate a range for the answer just by looking at the factors. They easily think about ten more than a number or three hundred less. This allows them to manipulate numbers to simplify calculations. This might look like turning $158+98$ into $156+100$, turning 360×5 into 3×5 and 6×5 and adjusting for place value, or simply moving fluidly between numbers knowing that 1,532, 1,632, 1,732 are each 100 more than the previous number.

DOI: 10.4324/9781003230915-18

As a former fifth grade teacher, I appreciate how strong number sense allows fifth graders to understand the relationship between digits in a number and, as a result, comprehend that when a number shifts place values, its value becomes ten times as much or 1/10 as much. For example, between 70,204 and 72,004, the 2 is worth ten times as much (200 versus 2,000 or conversely 1/10 as much). The bottom line is when students are comfortable with this base-ten system and have a developed number sense, they feel confident with numbers and can use their predictability to solve new and challenging problems. Memorization and algorithm-based calculation is less essential because they know how to manipulate numbers to work for them.

As we began the research for this Lesson Study cycle, I was excited to learn about what went into supporting this deep understanding and familiarity with base ten in first grade.

This Lesson Study took place at my former school Acorn Woodland Elementary School, located in East Oakland. Acorn has a high population of English language learners, and a majority of students qualify for free and reduced-price lunch. We were fortunate enough to have the opportunity to participate in a multiple-year Lesson Study grant that allowed us to work with Mills College. As part of this, we developed a research theme to guide our multiple Lesson Studies taking place each year. Teachers joined Lesson Studies both in and outside their own grade band to help them develop a more complete sense of the standard progressions and the math learning that took place in our school K–5.

As a school, we were in the middle of a multiple-year research theme thinking about how we can deepen student understanding through problem solving. We were aiming to use a problem-solving approach to teach math whenever possible. Our belief was that when students were given the chance to grapple with a "just right" problem and apply what they know to reach new, deeper understandings, the learning was also deeper and more meaningful. Memorization of ideas, algorithms, and tricks would not be as necessary if you can recreate them because you know how they are derived.

Therefore, the research lesson in this Lesson Study cycle was taught using a problem-solving approach. Students are given a problem they have never seen before, and use what they already know to solve it. Often, this problem requires students to use numbers in a new way to take their understanding a step further. After solving the problem, the teacher strategically asks a few students to share how they solved the problem, and the class discusses what they see. They compare and contrast ideas.

For this lesson, we were curious how students would use their knowledge of place value to support addition with two-digit numbers where regrouping is necessary. We wanted to see if they would see the utility of grouping ten ones as a unit of ten. As an upper grade teacher, I was excited to see how this very foundational understanding of place value was acquired, as the connections to math in the upper grades were clear.

The first graders were in the middle of a unit about place value. They had spent time counting objects by ones and by groups. They had counted by tens. When given a two-digit number, they had built it with manipulatives such as base-ten blocks and connecting cubes. The first graders also had exposure to a T-chart as a method for organizing tens and ones. They had many experiences with the concepts of addition and subtraction.

Students had also participated in daily number talks with ten frames, practicing manipulating two single-digit numbers to make a new ten with some ones (for example, when shown a 9 and a 5 on two ten frames, students could see that they could "make a ten" and change the problem to 10+4). Throughout these learning opportunities, students had shown varying levels of understanding of place value. We compiled the research we did during this Lesson Study cycle along with our unit and lesson plans into a Lesson Study discussion guide to be distributed to the

Figure 4.1 Base-ten blocks

Figure 4.2 Connecting cubes

rest of our colleagues on the day of the research lesson. As stated in our Lesson Study discussion guide:

> Our experience has shown us that students more readily see the number 38 as 30 and 8, versus 3 tens 8 ones. The challenge in our unit is to deepen student understanding of place value and help them to recognize and understand that a double digit number is composed of tens and ones. The intent in developing the unit progression has been to provide students the necessary learning experiences to be able to decompose any two-digit number into tens and ones, and then use that knowledge to add tens to tens and ones to ones.

Given these experiences, I wanted to see if first graders were ready to let go of counting by ones and to truly make use of the ten as a unit. Could they see that

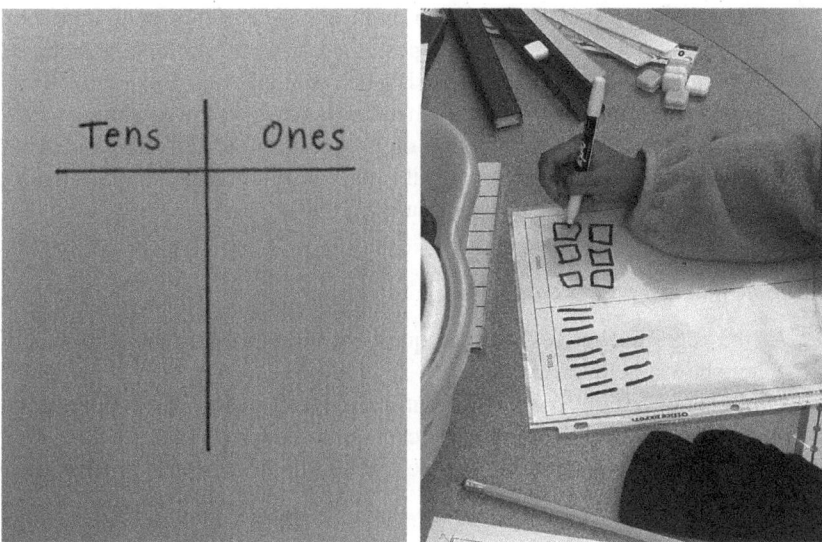

Figure 4.3 Tens and ones T-chart

using the ten was both more efficient and more accurate? Would they take the first steps on their own towards creating the "line up the tens and ones" algorithm students are so often taught to memorize by making use of base-ten manipulatives, T-charts, and their own problem-solving strategies?

Lesson plan

Title of the lesson: Counting Mr. Gaines' Stickers!

Learning objective: First graders will solve a two-digit addition problem with regrouping, applying what they know about place value to make sense of what to do with a new ten.

Students' school: Acorn Woodland Elementary School, Oakland, CA

Student ages: 6–7 years

Instructor: Evanne Ushman

Co-authors: Evanne Ushman, Julie Guy, Shelby Halela, Nakachi Clark-Kasimu, Ginger Cook, Jana Morse, Kari Laux

Date: March 22, 2017

Goal of the lesson: Students will see that when they add and there are more than ten ones, they can make a new ten.

Learning standard: CCSS.MATH.CONTENT.1.NBT.C.4: Add within 100, including adding a two-digit number and a one-digit number, and adding a two-digit number and a multiple of 10, using concrete models or drawings and strategies based on place value, properties of operations, and/or the relationship between addition and subtraction; relate the strategy to a written method and explain the reasoning used. Understand that in adding two-digit numbers, one adds tens and tens, ones and ones; and sometimes it is necessary to compose a ten (Common Core State Standards Initiative, 2010).

Introduction and posing the problem

At the rug, Teacher reviews the question and students read it together. Teacher shows them the stickers.

"Mr. Gaines bought stickers for our class! He bought 27 Minion stickers and 38 emoji stickers. How many stickers does he have in all?"

Students transition to desks and go through the three-reads process as a whole class.

a) First read: Context
b) Second read: Numbers
c) Third read: Looking for clue words on whether an addition or subtraction problem.

Teacher elicits equation from the students:

a) $27 + 38$

Students are given five to seven minutes to write their own ideas and solve.

Students may use manipulatives at the tables to make sense of the problem. They will have access to base-ten blocks.

Anticipated responses

R1: Student counts on from the larger (or possibly even the smaller) number. They will put one number in their "pocket" and then add the other number on by counting by ones.

R2: Student counts all the tens (50), then counts on by ones (15) for a total of 65.

R3: Student finds a new ten in the ones and adds it to the other tens for a total of 6 tens and 5 ones.

Comparing and discussing

The following are potential class discussion questions our team thought could prove helpful depending on what we saw from students.

"Student 2 added the stickers and got 5 tens and 15 ones and student 3 added the stickers and got 6 tens and 5 ones. . . . How can these both be 65!? These seem different! You have 6 tens; you have 5 tens. How can they both be 65?"

"Student 3 added the stickers and got 6 tens and 5 ones. We figured out that 6 tens and 5 ones is the same as 60 + 5. Wouldn't our answer then be 605?! Why is it 65?"

"What can we do if there are more than 10 ones?"

Summing up

The teacher will devise a summary statement with the help of the class. Generally, the summary is based on what has been discovered together as a math community. The teacher can assist with this through thoughtful questioning and facilitation. The hope for this summary statement today is that it will be something like this: when we add and there are more than ten ones, we can make a new ten.

Board plan

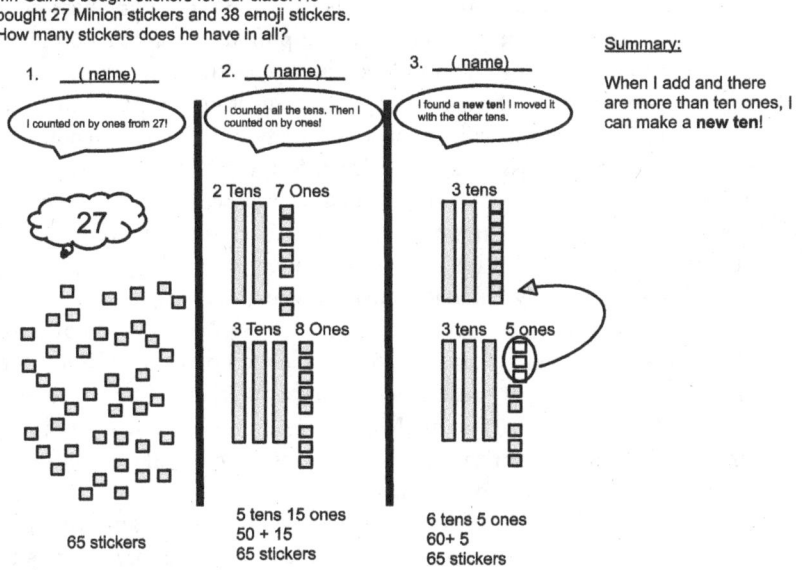

Mr. Gaines bought stickers for our class! He bought 27 Minion stickers and 38 emoji stickers. How many stickers does he have in all?

1. ___(name)___

I counted on by ones from 27!

27

65 stickers

2. ___(name)___

I counted all the tens. Then I counted on by ones!

2 Tens 7 Ones

3 Tens 8 Ones

5 tens 15 ones
50 + 15
65 stickers

3. ___(name)___

I found a new ten! I moved it with the other tens.

3 tens

3 tens 5 ones

6 tens 5 ones
60+ 5
65 stickers

Summary:

When I add and there are more than ten ones, I can make a **new ten!**

Figure 4.4 The board plan with anticipated student responses and the summary for this lesson

What I learned

It was clear to me that students needed many experiences with numbers to gain comfortability with base ten and to develop number sense. Number sense is not taught in a single moment, but acquired through repeated experiences of experimenting with and exploring numbers. It is built by talking about numbers with other mathematicians, tinkering with them, and coming to conclusions about how numbers work as students are ready for them. It comes from investigating what sometimes works, what doesn't work, and what will always work. It cannot be told to you or memorized. It has to be experienced and learned. Perhaps my best evidence of this is my own growing understanding of place value as we researched for this Lesson Study cycle. As an adult, I still feel like I am coming to new realizations about the base-ten number system.

In the end, the actual board from the lesson turned out to be rather similar to our team's board plan as written in the lesson plan. Making a detailed board plan before the lesson was a powerful exercise in anticipating student responses and preparing as teachers for how to monitor student work during the lesson and sequence the share for the discussion. Though a lot could be learned from the lesson itself and the discussion that took place on the carpet during the lesson, for me, the real learning was in the narrative that unfolded around the room during student work time and in the discussion we had as teachers after the lesson was over.

After observing the lesson, the adults in attendance gathered together as we always did after a research lesson to share what we had seen. Each person shared what they had observed happen at each table, helping us compile a complete picture of the classroom during student independent work time.

During independent work time, I saw the stepping stones of place value understanding. Students were on all different levels of how they were ready to think about place value, from most tangible to least. Some preferred to use manipulatives to count out each number; some needed to draw each item to know that it was there. Some could write a number, such as 2, 5, or 10, and circle it, and trust that that was indeed 10 even if they hadn't made ten lines. This was poignant for me. I know I have definitely made the mistake of using manipulatives or models as a formality, almost out of a need to "check the box" of having covered that. Seeing Evanne's students attack the math using the method they were comfortable with reminded me that all of these things – manipulatives, models, representations – should be in service of the math. They should help students grapple with new learnings, or explain their thinking. They should help make the abstract tangible, until students are ready for the abstract.

After reflecting on this lesson as a Lesson Study planning team, we realized it came too soon for students to grasp the intended learning outcome. Many students were still counting on, and many were miscalculating. Surprisingly, many students were not yet ready to see ten ones as one ten. We had painstakingly planned this math unit, Evanne had taught, we had come back together and recalibrated and continued on. All of this was in hopes that this research lesson's topic would be presented at the right time for students to grapple with it, that this problem would feel "just right" when we presented it. We were wrong. What we saw from students on the day of the lesson made it clear that we needed to give these young mathematicians more time with these concepts. Conceptual understanding takes time, and we had rushed things. In his final comments, Dr. Takahashi suggested a progression for this unit that was a bit slower. He suggested gradually working up to a problem like ours ($27 + 38$) by taking the following steps:

Table 4.1 Possible progression for additional problems in this unit

Addition problem	Reasoning
$30 + 7$	Help students move away from counting on. Spend some time here allowing students to soak in this conceptual development and then develop procedural fluency. Give similar problems where the ones can be put in the ones place.
$30 + 20$	Students realize they can simply add the tens. Reinforce the idea they do not have to count on.
$32 + 24$	Students add without regrouping. They see that they can add tens and tens and ones and ones – i.e. $(30 + 20) + (2 + 4)$.
$30 + 24$	Allow students to grapple with zeros. These can be tricky.
$32 + 4$	Students see what happens when there are no tens to add.
$27 + 38$	Students try regrouping. They hopefully see that ten ones are also one ten.

As Dr. Takahashi explained this possible progression, he highlighted for us that building conceptual development needs to be thoughtful. This thoughtfulness can sometimes look like choosing to pause. After trying $32 + 24$ for example, it might be necessary to pause for a day or two to give this idea of adding tens and ones separately time to soak in. Dr. Takahashi likened this to building a physical structure. At certain points, you may need to wait for things to dry.

Beyond learning more about place value, Lesson Study (both this one and in general) taught me a lot about what kind of conditions create the best math learning opportunities for students. I learned that students must see themselves as mathematicians. What I saw in Evanne's classroom was a culture that did not happen overnight. It is a culture I saw growing throughout our school as we continued to apply a teaching-through-problem-solving approach to learning math. When students identified as mathematicians, they were willing to take risks to solve a problem, even if they had never seen that type of problem before. They trusted in the math they knew and in themselves as mathematicians to experiment in their problem solving. They tinkered, using what they knew to go deeper. They had agency and a high level of choice in the math classroom. It was their math. They did not wait for a teacher to come tell them how to do it or correct their steps. On their path to understanding, they built off what they knew and reflected as they went.

This student reflection is key. After the class discussion in Evanne's classroom, students were asked to turn in their notebooks. Dr. Takahashi pointed out that we had missed a key opportunity after the discussion to have students reflect on their learning and update their thinking. In his final comments, he talked about how important it is that each student walk away from the lesson feeling successful. Additionally, he described how important it was that students could then update their work at the end of the class discussion to reflect their new understandings and, hopefully, more accurate calculations. While students might need more time to internalize the concept for themselves, regardless, time should be taken at the end of each lesson to reflect.

Having tried this regularly in my own classroom, I can attest to the power of reflection. It surprised me when I realized how rare it was that we ask students "What did you learn?" after a lesson and give them an open runway to answer. As teachers, we may finish a lesson feeling confident students learned one thing, but when we ask them to reflect openly, their responses may surprise us and will likely allow us to be more responsive teachers.

Furthermore, I realized that so much of what I see in the upper grades, when students lack number sense or seem to be guessing without thinking deeply, is from many years of lacking agency and identity as mathematicians. Students who have approached math learning as concepts to be memorized and regurgitated only know what to do when the question presented is close enough to what they have previously seen. They don't know how to build and extend their own learning. Instead, the foundational understandings of base ten have to be grappled with. A child has to decide that they believe that ten ones make a ten and that it is useful to see them as such. If they are simply told, it's almost as if they don't trust it. They won't use it as flexibly because they are just using what they have memorized, not what they understand.

This grappling produces the types of mathematicians we strive to foster as teachers. Mathematicians who have convinced themselves and built their own understanding can solve problems with situations or number types they have never seen before. They can explain their reasoning. They can even unpack the misconceptions of their peers. It results in a classroom that is buzzing with math learning. It results in a classroom where students teach students and everyone is a valuable member and contributor to the learning.

Through the research for this and other Lesson Studies, my own understanding of elementary mathematics grew immensely. I found that as I improved my content knowledge, and specifically my knowledge of the standard progressions, I became a more effective teacher. The more I learn about the math, the more I am able to recognize partial understandings in my students. What I would have previously written off as simply incorrect understandings or misconceptions about numbers are now seen as opportunities and stepping stones to deeper understanding. Additionally, the more I learn to question student thinking and dig deeper into their ideas, the more I am able to approach my teaching with an asset-based lens, helping students build on what they know to create new understandings. The reality is that just because students have seen or been directly taught the math doesn't mean they "get" the math. New math understandings are not achieved all at once.

Over time, my role as a teacher has shifted greatly. Before, I often played the role of standing before the class, saying, "This is how you do this." Now I am a facilitator. My job is to provide intentional problems that allow students to grapple with the math and to ask thoughtful questions that help them compare methods and have deeper conversations between students. My job is to see what kids know how to do, and connect them to the next idea they are ready for. So much of this relies on the teacher having a robust knowledge of mathematical concepts they teach. One way to gain this is through Lesson Study. Pursuing this robust under-standing is a life-long journey.

Above all, through observing Evanne's lesson and my general participation in the Lesson Study process, I have learned that student math learning is a contin-uum. There is no magic bullet or perfect explanation that can help a student inter-nalize the math concepts. It takes experimentation with the numbers, struggle, and many opportunities to discuss with peers and observe their thinking to ultimately learn new math ideas. Our job as teachers is to facilitate these opportunities, to observe, to ask thoughtful questions, and, often, to get out of the way.

Reference

Common Core State Standards Initiative. (2010). *Common Core State Standards for Mathematics*. www.corestandards.org/Math

Discovering numbers greater than 1000

Making meaning of numbers above 1000, for second grade (7- and 8-year-old) students

Rebecca Reddicliffe

The focus of this section is a research lesson taught at Chavez Multicultural Academic Center, a public Kindergarten-through-eighth-grade elementary school in the Back of the Yards neighborhood in Chicago. At this time, the teachers participating in Lesson Study developed a school-wide research theme. Our research theme was "students will construct viable arguments and critique the reasoning of others." We wanted students to be able to use mathematical language to describe and justify their thinking, as well as respond to, add on to, and critique the strategies of their peers. The research theme for the unit and lesson came from Common Core State Standard for Mathematical Practice 3: construct viable arguments and critique the reasoning of others (Common Core State Standards Initiative, 2010). Our school-wide research theme was chosen because we wanted to focus on students developing the skills to defend their math conclusions, discuss their math with peers, and question or critique their peers' math and their own math. We saw that there was a great need for students to improve their ability to explain their reasoning using mathematical language or models. Through math journals and problem-solving lessons, we expected that students would improve in providing justification for their own thinking, and be able to question, critique, and respond to their peers' math and reflect on their own thinking. We wanted to further expand and develop our ability to facilitate mathematical discussion among our students. This would ultimately achieve our goals of helping students verbalize their ideas, develop mathematical vocabulary, and foster a trusting learning community.

As the teacher of this research lesson, I learned an immense amount about how my students best learned a new concept. This lesson reminded me of the importance of patience and struggle, and it emphasized that students learn a concept best when the knowledge comes from themselves and is made sense of with peers. When students have the opportunity to justify their mathematical reasoning to their peers and are given space to reflect on their own ideas and the ideas of their peers, they develop a deep understanding of the mathematical concept.

When the second-grade team met to plan this research lesson, we had just begun using a new math curriculum. We noticed a unit exploring numbers greater than 1000. This unit topic is not technically a second grade Common Core standard; it is, instead, a third grade standard. Still, we chose to teach it toward the end

DOI: 10.4324/9781003230915-19

of the year. We chose this topic for a few reasons. We noticed that our students had a solid understanding of place value. As we used our new math curriculum, students developed a strong sense of the base-ten system. They were so excited when we learned about numbers greater than 100, and since that moment, I had students asking when we would learn about numbers greater than 1000. I had also noticed students struggling with how to say numbers when we read, as well as students struggling to conceptualize a quantity if it was over 1000. As a second-grade team, we also observed that some students got stuck or confused when counting from one place value into another – for example, 1 more than 999 or 1 less than 1000. Furthermore, when students estimated a larger quantity, they tended to make illogical estimates. There may have been 900 of something, but some students would say the largest number they can think of. We thought this unit would improve our students' understanding about the relative size of numbers. I knew students would be eager to extend their understanding of place value. We as a second grade team had never taught this lesson before, and we hoped it would really allow us to evaluate our students' understanding of the base-10 system and their ability to manipulate numbers according to the rules of that system.

Most of the students at Chavez are English Learners, and this gave students the opportunities to use academic vocabulary to discuss numbers greater than 1000. It is also one of the last times students would see a new place value concretely, and we hoped our EL students (and all of our students) would gain a lot from the visual representation of numbers greater than 1000. The unit gave students a chance to manipulate place value cards to deepen their understanding of different ways to express and make numbers. Another factor in choosing this unit was that our students would be able to express numbers greater than 1000 in words and numerically. We were optimistic that seeing the connection between these two ways of expressing a number would be very beneficial for our second graders.

In this research lesson, students were presented with the task of counting a large number of pennies. We had not formally counted above 1000 before. Here is the description of the research lesson:

> Students will receive a picture with 2,354 pennies. Students will grapple with how to count, name and write a number that is greater than a thousand. Students will recall that when they previously counted large numbers of objects in previous lessons, they made groups of tens and hundreds. Students will build off their knowledge from a previous lesson to see the efficiency in counting in groups, and even counting with groups of a thousand.

Why did we decide on this research lesson?
Our research plan explains:

> Although CCSS (Common Core State Standards Initiative, 2010) specifies that 2nd graders only name, read and write numbers up to 1000, the Japanese curriculum we are using justifies students learning numbers up to 10,000 as it shows students the continuity of the base-10 structure. We wanted students to see the connection between the regularity of the base-10 structure of 4- and

5-digit numbers and that of 3-digit numbers learned previously. Furthermore, we hoped that students would see that 10 groups of the previous place value will make a unit that is one place value greater. This lesson will lead them to make that connection, as they will see the picture of the pennies arranged in multiple ten by ten squares. We chose to present the pennies this way to encourage students to see the base-ten structure of a number greater than 1000. If we give students 2,354 pennies, the regularity of the base-ten structure would be less clear, and counting would certainly be more frustrating.

The goal in 2nd grade is also to have students continue developing a good number sense and seeing numbers in multiple different ways. Students will be able to read the number 3240, represent it as 3000 + 200 + 40, and describe it as 3 thousands, 2 hundreds, and 4 tens. They draw on that knowledge to figure out which representation is best to learn new concepts and also understand the relative size of numbers.

2nd grade is the last time students will see a visual representation of a new place value. Therefore, it is important for students to have a good command of the concept of the base-10 system as they work theoretically with place value.

3rd grade students will be expected to read, write, and reason with numbers up to and including the ten millions place. This is not a 3rd grade Common Core standard; the standard is actually read, write, and reason with numbers up to 10,000. In 1st grade and 2nd grade, the curriculum, which follows a traditional Japanese approach, slowly introduces the place values of tens, hundreds, and thousands to give them the foundation to fluently reason with 5 additional place values. Students can generalize their understanding that "10 groups of a place value create another larger unit" for the place values of hundred thousands, one million, and ten million. Students should also be able to recognize that the "1" in the number 3010279 represents not "1" of something, but 1 ten thousand of something.

This lesson taught me that 7–8-year-olds, when given the mathematical background necessary, can access numbers greater than 1000. Students with a strong foundation in the base-ten system and the ability to problem solve were able to count above 1000 for the first time. This research lesson reminded me how important it is to be patient and let students struggle with, reason about, and discuss new mathematics with each other. When given the opportunity, all students can be successful in math.

Lesson plan (shortened)

Title of the lesson: Let's Investigate Numbers Greater than 1000
Students' school: Chavez Multicultural Academic Center, Chicago, Illinois
Student ages: 7–8 years

Instructor: Rebecca Reddicliffe
Co-authors: Bethany Jorgensen, Ana Cabrera, Adrianna Soto
Date: May 18, 2018
Goal of the lesson: Students will deepen their understanding of numbers as they expand their knowledge of the base-10 structure of numbers up to ten thousand. Students will recognize the benefit of making groups of tens and hundreds to count, connect grouping of tens and hundreds with the base-10 numeration system, and use base-10 positional regularity in the process of calculating. Students reason about and express the base-10 numeration system, demonstrating an understanding of the relative size of numbers. Students read and write four-digit numbers and express the relative size of numbers using symbols for equality and inequality. Students will understand the structure of numbers, know how to read and write four-digit numbers, and compare the relative size and order of numbers.

Learning standard

CCSS.MATH.CONTENT.2.NBT.A.1: Understand that the three digits of a three-digit number represent amounts of hundreds, tens, and ones; e.g., 706 equals 7 hundred, 0 tens, and 6 ones. Understand the following as special cases: 100 can be thought of as a bundle of ten tens – called a "hundred" and the numbers 100, 200, 300, 400, 500, 600, 700, 800, 900 refer to one, two, three, four, five, six, seven, eight, or nine hundreds (and 0 tens and 0 ones).

CCSS.MATH.CONTENT.2.NBT.A.2: Count within 1000; skip-count by 5s, 10s, and 100s.

CCSS.MATH.CONTENT.2.NBT.A.3: Read and write numbers to 1000 using base-ten numerals, number names, and expanded form.

CCSS.MATH.CONTENT.2.NBT.A.4: Compare two three-digit numbers based on meanings of the hundreds, tens, and ones digits, using >, =, and < symbols to record the results of comparisons (Common Core State Standards Initiative, 2010).

Lesson flow

Introduction

Presentation of ideas

a   Our presentation of ideas starts with giving our students a familiar context. Our school has a fundraiser called Pennies for Patients. This context provides a reason for students to count the pennies.

b   We were worried that some of our students would want to count the dollar amount of the pennies, as this is what many of them have done with money (and specifically pennies) in a past lesson. Because of this, we thought stating the number of pennies the second grade classes collected would steer students away from counting the dollar amount and toward counting the number of pennies.

c   For the estimation, we will show our students an enlarged color copy of the pennies. We don't want them to be able to start counting themselves in this part, so we decided to have them all look at the picture together.

Posing the problem

The task is in a textbook that follows a typical Japanese approach. My students are very familiar with this text, and we use it weekly. The issue is that the pennies are split between two pages. Students would be distracted by the book closing or trying to push down the pages. Furthermore, it automatically split the pennies into two separate groups that are not meaningful. For these reasons, each student will get a paper copy of the pennies. In order for students to group the pennies, they will also be provided with different-colored pencils or highlighters to use.

Students study the picture of many pennies and discuss the situation.

Remind students about the Pennies for Patients fundraiser we had in the winter at Chavez.

Ms. Corona is counting the pennies.

Ms. Soto's class collected 649 pennies. Ms. Jorgensen's class collected 806 pennies. Our class collected 917 pennies.

Now Ms. Corona is counting Ms. Cabrera's pennies. She is tired of counting, so I told her we were excellent counters. She needs our help!

Look at the pennies. How many pennies do you think there are? Why do you think so?

– Students estimate the number of pennies by looking at the picture.

Anticipated responses for estimation

– One group of pennies has 100 pennies, so there are more than 100 pennies.
– There are many groups of 100 pennies, so there are 1000 pennies.
– One group of pennies has 100 pennies, and there are more than 10 groups, so there are more than 1000 pennies.

Hatsumon: How many pennies are there? Think about the way you're counting the pennies.

Anticipated responses

R1: Circle 100 pennies as a group and use the grouping strategy to count the total.

Date;

How many pennies are there?

Think about the best way to count
the pennies.

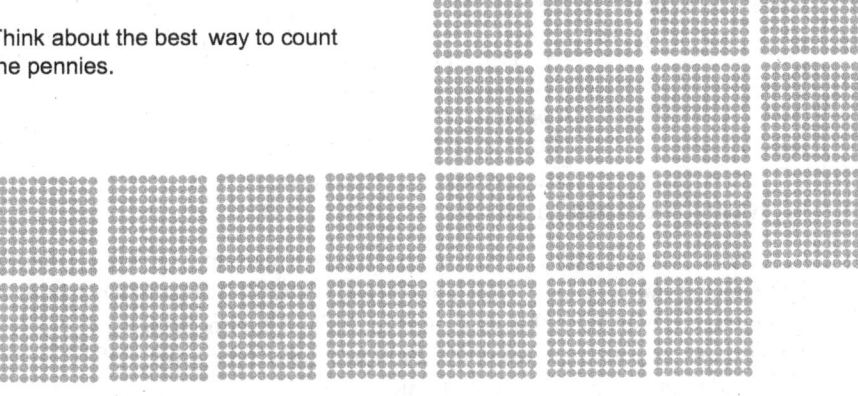

Figure 4.5 A copy of the task students received

- 23 groups of 100 and then 5 tens and 4 ones.
- 23 hundred 54

R2: When counting up to 1000, use a different color (or different circle) to
show groups of a thousand.
- 2 thousands, 3 hundreds, 5 tens, 4 ones
- Get stuck after 1000
- $1000 + 1000 + 300 + 50 + 4$
- 2354 (twenty-three fifty-four)
- twenty-three hundred fifty-four

Table 4.2 A chart students may create as a way to organize their counting

Hundreds	Tens	Ones
23	5	4

Comparing and discussing

a   Students will turn and talk to their math partner to find out how their partner counted. Students will explain to each other how they counted the pennies. This will give us insight into students' thinking and will also allow the
teacher to circulate the room to hear students' strategies.

b   During our discussion, students who come up to share their ideas will explain
how they counted the pennies. Because so many of our students are ELLs, we
decided each student idea would be accompanied by the picture of the pennies. Students will explain how they counted, and the teacher will circle the
pennies according to what the student says.

Comparing and discussing

a   I circled a group of 100 pennies and saw that there were 23 groups of 100 and
    then 54 more pennies, but I don't know how to say the number of coins.
b   Place value chart

Table 4.3 Place value chart

Hundreds	Tens	Ones
23	5	4

b2. twenty-three hundred fifty-four

c   I circled a group of 100 pennies. When I made 10 groups, I made a group of
    1000 pennies and circled it with a big circle. I noticed there are two groups of
    1000, three groups of 100, five groups of 10, and 4 more pennies.
d   Two groups of 1000 pennies make 2000 pennies. Three groups of 100 make
    300. And then we have 54 more pennies.

How many groups of 100 pennies are there?
How many groups of 1000 pennies are there? Why do you think we made groups
    of a thousand?

During the comparison and discussion, the teacher may prompt students in the
following ways:
Help students recognize there are 23 groups of 100 pennies. (Students recognize this by listening to each other.)
Remind students how we regrouped if there were more than 10 tens or 10 ones
in the place value chart.
Help students recognize that when they make a group of a thousand, they do
not need to go back and count again the number of hundreds. Also, it is easier to
see the number by having two groups of 1000 instead of having 20 groups of 100.
Help students recognize there are 2 groups of 1000. Make sure all students see
and circle the groups of 1000.
If students do the place value chart, ask if we've ever put two digits in one
column – if there are more than ten 10s or 10 ones, we regroup. . .

Summing up

a   Students will turn and talk to their partner about what they learned today. This
    is a good way for the teacher to begin to evaluate what students took away
    from the lesson.
b   The class will develop a summary statement for our learning.
c   Reflection: Students will reflect on what they individually learned or noticed
    from the lesson. This gives them time to reflect on their own learning and

gives the teacher a way to understand where students are relative to the lesson goal. Students are provided sentence stems for their reflection.

We learned that ten groups of 100 make "one thousand" and two groups of 1000 make "two thousand." We can use groups of 1000 to count more quickly.

(Even though this is not our main goal, we feel that some students will view counting by using groups of 1000 as their main learning)

OR

Main Goal: We learned that the number made of two 1000s, three 100s, five 10s, and four 1s is two thousand, three hundred fifty-four.

Board plan

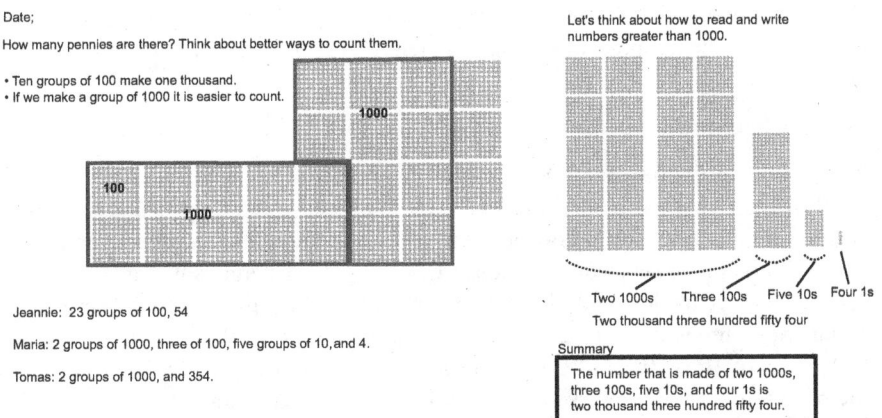

Figure 4.6 The board plan

Observations from the lesson

After the initial problem-solving time, there was not a single student who had counted the pennies exactly, but there were many students who counted in groups of 100, and a few students who found 2 groups of a thousand. One student counted by circling the groups of 100 and then counting 5 tens and 4 ones. Another student counted two groups of a thousand, but struggled with how to count on from 2000, so he left out the 3 hundreds, 5 tens, and 4 ones. After both of those students presented their thinking, students were able to discuss the best way to count, and why students had difficulty counting the total number of pennies. By the end of the lesson, the class summarized their learning in the following way: today we learned we can make groups of one thousand, and we made the number two thousand, three hundred fifty-four.

One point from the post-lesson discussion was that the discussion could have focused more on why counting past 1000 presented difficulty, and upon reflection, I agree. If we had emphasized the reason it was difficult, students may have been better able to see the benefit of counting in groups of 1000. Still, the student presentation of ideas showed the students how counting by 1000 made finding the total number of pennies more efficient.

There was a large discussion of whether or not the teacher should have written the number of pennies as a four-digit number (2354) in addition to writing 2 thousands, 3 hundreds, 5 tens, and 4 ones. Ultimately, the teacher chose not to. A student did ask how to write it, and others were clearly grappling with that, as well. At the time, though, it seemed that the most important new learning was not how to write a four-digit number, but instead that 2 thousands, 3 hundreds, 5 tens, and 4 ones make two thousand three hundred and fifty-four. In the final comments, Dr. Takahashi agreed, and emphasized the importance of patience. As we reflected on this lesson, we realized how tempting it can be to rush students along, tell them how to count the pennies, or just say to them that counting by thousands is better.

What I learned

This lesson taught me that 7–8-year-olds can readily access numbers greater than 1000, it reminded me of the importance of patience and struggle, and it emphasized that students learn a concept best when the knowledge comes from themselves and is made sense of with peers. Counting 2354 pennies was a challenging task for 7–8-year-olds, and all students struggled. Still, they were determined to count the pennies and make sense of the large quantity. Whether they counted by groups of 10 or 100 or 1000, all relied on prior knowledge of the structure of the base-ten system to work toward gaining a new understanding. Their struggle was productive, and even though not a single student could verbalize the exact number of pennies on their own, they arrived at the correct number of pennies together. They admitted their confusion or mistakes and turned to their peers, not their teacher, for guidance. This lesson was powerful for me because it truly highlighted how powerful learning can be when students are given the support they need to access new content and the opportunity to productively struggle with and learn from their classmates.

Impacts on my own teaching

This lesson, and my work with Lesson Study more broadly, influenced everything I do in my classroom. I learned not only that 7–8-year-olds can make sense of numbers greater than 1000, but also that they are capable of accessing new and challenging concepts in all content areas. After teaching this lesson, I was struck by how deep students' understanding of their learning was. Their productive struggle and the discussion with their peers led them to have a strong understanding of the new place value of the thousands place. They had ownership of their learning and felt successful. When students explore new concepts and productively

The Lesson Study
By Kenneth Herrera

Do you want to know about a class from Back of the Yards? The class is smart enough to do a lesson study. Come to find out how they crushed it!

I was getting ready to do a lesson study with Ms. Red. I was nervous and we got to the building. We got out of the bus. We went inside and sat down on our chairs.

We waited for a while and finally we were ready for the lesson study. We went to our spot on tables. I was near Armando, my friend.

Armando was the first one to get called, I disagreed. Then, another person got called, I also disagreed. I got called! I did my work. The number was five thousand three hundred fifty-four and the class agreed with me but Armando didn't agree. I called on him and convinced Armando that how you call it!

My class and I are clever and don't say anything that is not true. Let the people in the world hear our story. A lesson study is important because it can help you learn. Also, a lesson study lets all students show what they know!

Figure 4.7 A student reflects on his experience as a participant of a research lesson. He wrote this article for his third grade class's magazine titled "Say It Loud."

struggle independently and with peers, their learning is deeper and more powerful. They are the creators of their new knowledge. It seemed so simple, but I had only provided students with this opportunity in math. Now, every time I sit down to plan a unit in any subject area, I ask myself, "How can I facilitate this lesson or unit so that students feel ownership of their new learning?" This lesson wasn't just powerful for me. It also impacted the students. Over a year later, a student wrote an article about his experience for a magazine his third grade class was working on. As you can see in Figure 4.7, he may not remember the exact number of pennies, but he remembered that he and his classmates worked together to understand numbers greater than 1000, and that this lesson let "all students show what they know."

Reference

Common Core State Standards Initiative. (2010). *Common Core State Standards for Mathematics*. www.corestandards.org/Math/

Student learning: addition with fractions

Adding fractions with like denominators, grade 3 (8- and 9-year-old) students

Lindsay Singer Kalt

During my career as a third grade mathematics instructor, I have grappled with the question of how to best introduce addition of fractions to my students. Over the years, in my own classroom and through observation of others, I have witnessed frustration from students and teachers alike stemming from how to best teach/learn the standards focused on fractional numbers.

In second grade, the United States Common Core State Standards (Common Core State Standards Initiative, 2010) asks that students learn how to partition rectangles and circles into halves, thirds and fourths.

- CCSS.Math.Content.2.G.A.3: Partition circles and rectangles into two, three, or four equal shares, describe the shares using the words halves, thirds, half of, a third of, etc, and describe the whole has two halves three thirds and four fourths (Common Core State Standards Initiative, 2010).

Students are asked to have a concrete understanding of what it means to partition shapes into equal parts. In grade 2, students do not yet think about fractions on a number line or naming fractions using numerical representation. Yet, the Common Core State Standards for third and fourth grade ask students to drastically widen their understanding of fractions. During these grades, students focus on fractions on a number line, equivalent fractions, comparing fractions and also adding and subtracting fractions.

According to the Common Core State Standards in Mathematics, students should master the foundational concepts of fractions in third grade. Essential to this understanding are the dual concepts of partitioning and iterating. At Chavez, teachers found that there were a plethora of misconceptions in regards to fractions. Before determining which lesson in the unit to focus on, we first focused on the most common misconceptions we see in the classroom and how problem solving–based teaching can help to correct or avoid these issues.

1 Students see the numerator and denominator as separate values, and it is hard for them to see ¾ as one number.
2 Students think all fractions relate to ratios and probability, again seeing the fraction as two separate numbers.

DOI: 10.4324/9781003230915-20

3   Students think that fractions can be any parts, not equal-sized parts.
4   Students do not see fractions as numbers that can be placed on a number line between 0 and 1 and so on.
5   Students do not understand that all fractions can be simply understood as a collection of unit fractions.

Although most students understood how to add and subtract fractions with the same denominators, they could not vocalize why the process was mathematically sound. In short, students were able to follow a process but struggled to make viable mathematical arguments and use mathematical models for justification. It was for this reason that our team decided to focus our research lesson on the adding of fractions with like denominators. We wanted to make sure that we were teaching this concept in a way that gives students the ability to grow both their concrete and abstract understanding. Additionally, this lesson and students invented strategies for solving would give teachers a better grasp of students' understanding of the concepts that came before this lesson.

After doing our research, we determined that we wanted to focus on the addition of fractions lessons from Sansu Math. According to Sansu Math (Koyo Publishing, 2015), Chapter 14, the goals of the unit are that students will understand the meanings of fractions and how to express them. Throughout the unit, students will realize the benefits of using fractions to express the sizes of evenly divided parts that cannot be expressed as whole numbers as well as how they can use fractions in everyday life. Sansu Math directly addresses the misconceptions students can have in relation to the meaning of fractions, how to name them and what they are used for. In third grade, students need to learn to see fractions as iterations of unit fractions, first through a tape diagram and then transitioning to locating fractions on a number line.

The unit is designed to give students numerous opportunities to problem solve on their own and in collaboration with other students. Each day, students come up with mathematical solutions to problems that will add to their conceptual understanding of fractions. Students will work independently, but they will also spend time sharing their thinking and defending their mathematical arguments during whole-group math discussions. Students will take part in critiquing their work and the work of their peers. Students at Chavez are majority English Language Learners, and therefore the opportunity to not only do math but to talk about math is extremely important to their ability to problem solve. To assist, our students have sentence stems to guide them to be able to explain their thinking and prepare them to critique the thinking of others. Through these problem-solving experiences, the hope is that students are able to work through misconceptions and gain a better understanding of the concept. By focusing on addition of fractions with like denominators as our research lesson, the team was looking to test our theory on the most effective ways to introduce the computation of fractions and to gain insight into student thinking. We included the following into our lesson plan so that observers of the research plan would understand our idea:

As with whole number computation, providing tasks without rules will help students have a deep conceptual understanding. Invented strategies are important to their understanding of fractions, as they tend to use the students' number sense. It pushes students to recognize fractions as a value between zero and one.

It is important to note, that this lesson was the first time computation of fractions has ever been introduced. Students have had experience adding and subtracting whole numbers and decimals, but never fractions. Through the lesson study process, we learned that students need a strong understanding of fractions as iterations of unit fractions to be able to determine a strategy for adding fractions with like denominators. It is also important for students to have experiences with different models and manipulatives, with an emphasis on tape diagrams and number lines.

Lesson plan

Title of the lesson: Addition of Fractions

Students' school: Cesar Chavez Multicultural Academic Center, Chicago, Illinois

Student ages: 8–9 years old

Instructor: Lindsay Singer Kalt

Co-authors: Lindsay Singer Kalt, Mayra Velasco, Danny Kim, Marcella Cadena

Date: March 6th, 2019

Goal of the lesson: Students can add fractions with like denominators by using visual fraction models to explain their reasoning.

Learning standard: Students understand addition and subtraction of fractions as joining and separating parts referring to the same whole.

Bilingual considerations: A majority of the students at Chavez are considered English Language Learners, and there are a variety of supports in place to help them to access, reference and retain information throughout the lesson. All mathematical vocabulary is written in both English and Spanish to allow students to make language connections. Additionally, the problem is posed in English and Spanish so that students can focus on understanding and modeling mathematics. Visuals and pictorial demonstrations are used when available to increase all students' concrete understanding. Visual cues/physical gestures also student understanding of academic concepts of procedural instructionals. Lastly, mathematical sentence stems are taught and posted around the room so that mathematical writing and conversation can take place.

Lesson flow

Introduction

Ms. Singer's class is going to have a party! We want to make a new yummy drink to go with our cupcakes. Ms. Singer has 3/10 liters of fruit punch and 2/10 liters of lemonade. What would happen if we combined the amount? What kind of number sentence can we use to find the combined amount of fruit punch and lemonade?

Posing the problem

Once students agree that they need to use addition to solve the problem, the teacher will write the number sentence on the board. **3/10 L + 2/10 L = ?**

Hatsumon: Let's think about how to add 3/10 liters + 2/10 liters.

Anticipated responses

R1 (invented strategy): Students use a tape diagram and label each unit fraction. They then shade 3/10 and 2/10 and count up how many 1/10s they have all together (5/10 L).

R2 (invented strategy): Students create a number line from 0 to 1. They use the number line to "hop" 3/10 and then 2/10 to add up to 5/10 L.

R3 (invented strategy): Students decompose 3/10 and 2/10 into unit fractions to come up with the number sentence 1/10 + 1/10 + 1/10 + 1/10 + 1/10 = 5/10 L.

R4 (misconception): Students add the numerator separately from the denominator 3/10 + 2/10 = 5/20 L.

Comparing and discussing

R4 (misconception): ONLY use this if a majority of students are adding the numerator and the denominator separately. We will put this solution up first and use the following invented strategies to disprove the idea.

R1 will be shared first because it's the pictorial representation on a tape diagram and is the most concrete. It also shows each unit fraction.

- How did Student A add the fractions?
- Do you agree with his/her idea?
- How does this prove that 3/10 L + 2/10 L = 5/10 L?

R2 will be shared next because it represents fractions on a number line, which helps students understand that when adding fractions we are moving farther away from zero. Additionally, it also focuses on fractions as iterations of unit fractions.

- How did student B solve this problem?

- How is this solution /strategy similar to student A?
- How does this model look similar or different from the first model?

R3 will be shared last because it is related to both R1 and R2 and it is the visual representation of the summary for today's lesson.

- How did Student C solve this problem?
- How is this strategy similar to student A and B?
- How did this student use unit fractions to solve?
- How can unit fractions help us add fractions with the same denominators?

Summing up

Mathematicians add fractions with the same denominator by adding or counting up the number of unit fractions.

Board plan

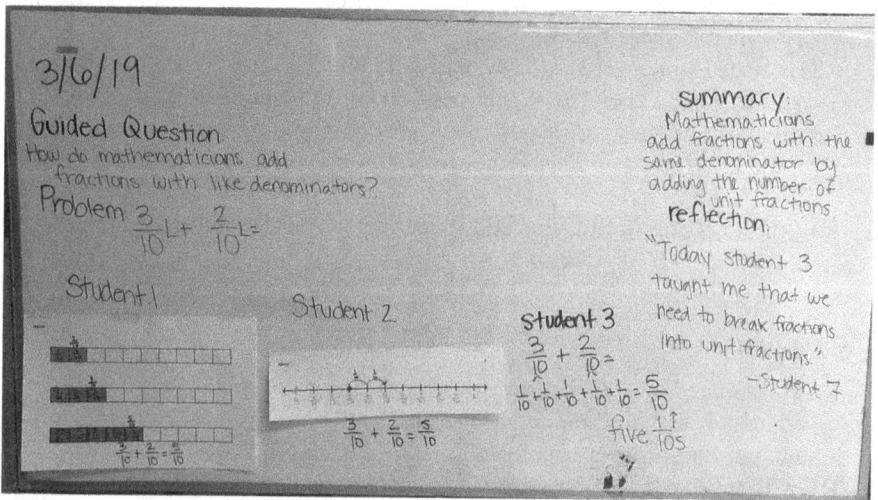

Figure 4.8 The board plan

Observations from the lesson

During the lesson, it became clear that students demonstrated a strong under-standing of unit fractions and also how to use models to help them solve. Most students used either a tape diagram or a number line to show their thinking. Early on, students demonstrated their mastery of fractions as iterations of unit fractions and also their comfortability using models like the number line and tape diagram. Lessons earlier in the curriculum we use gave students the knowledge they needed to be successful with this new problem.

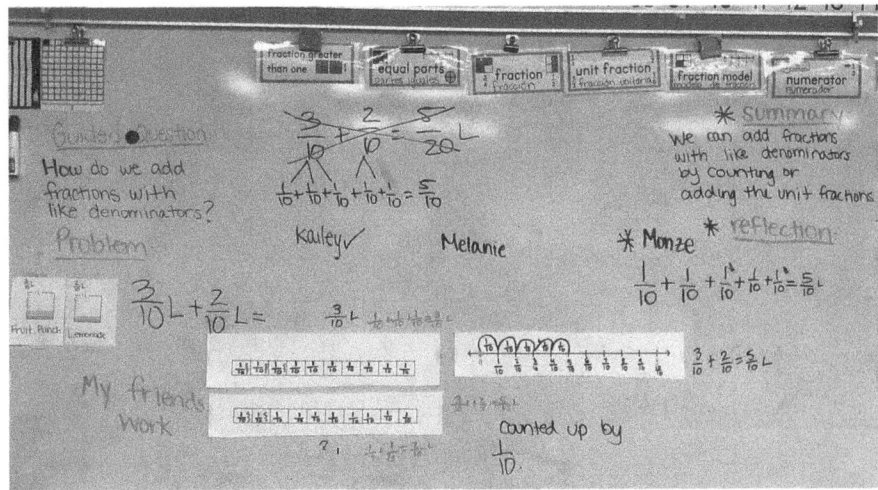

Figure 4.9 The board at the end of the lesson

Most students either opted to use the tape diagram or the number line to solve. A few students opted to use a number sentence where they first broke each fraction into unit fractions and then created a number sentence from there. These students also relied on their knowledge of unit fractions to solve the problem. What was most interesting to me was that there were about six to eight students who actually solved the addition number sentence first and then went back and created a model to prove that their answer was correct. Instead of using the model to solve the problem, they used it as evidence that their solution was correct.

When students came back together for the independent problem-solving discussion, they were able to explain their models to other students and defend their own solutions using mathematical evidence. Additionally, students were able to find similarities between the different solutions that were posted on the board. In my classroom, I ask students to write out their own solutions and lead the conversation. During this time, I stay towards the back of the classroom to position myself as the facilitator of the discussion. I want my students to do the heaving lifting of explaining their solutions, asking questions and critiquing the work of others.

Based on observations from the lesson and student math work in their notebooks, there was a plethora of evidence that students were able to come up with their own inventive strategies. By using student work and class discussion to make comparisons on the different strategies, students also determined that "when mathematicians add fractions with like denominators, they add or count up the unit fractions".

Figure 4.10 Students watch as Student C explains her solution at the board

What I learned/impact on teaching

After the lesson study process, I had some valuable takeaways from both the lesson and the post discussion that have helped to shape how I strive to teach fractions now and in the future.

Number lines: When teaching fractions in third grade, it is important for students to have a complete understanding of the number line. When we compare whole numbers, we know a number is less because it is closer to zero. This is an understanding that students already have. When teaching fractions, we need to continue this learning by bringing this understanding to the forefront. As teachers, we should be using number lines most frequently in models, so that students see fractions as different distances from zero. This will help students not only as they start to add and subtract fractions but also as they continue their learning with multiplication and division of fractions as well. Additionally, number lines are a more usable tool when it comes to comparing fractions and finding equivalencies. My first year teaching fractions, I was using a mix of representations, including pie diagrams. While this is not necessarily harmful, we do know that when we focus too much of our teaching on fractional parts of a circle (pizza fractions), students start to represent fractions only in this way – when in reality, fractions can represent numbers, distances and areas of all shapes. Fraction pies specifically are extremely hard to partition into equal parts, and they can confuse students as they are comparing and finding equivalencies. Even more important, if the circles are partitioned incorrectly, students are not actually representing fractions. In my own classroom, I have found that students find the number line to be a much more effective tool when solving and reasoning.

Modeling for reasoning vs. modeling to solve the problem: From the data collected during this research lesson, it was clear that some students were using the model to prove their answer was correct while others were actually using the model to solve. Students who were using the models to reason mathematically had a more abstract understanding of what was happening when we add fractions with like denominators. Although they did not need the model to solve, they used it to explain their hypothesis that they were adding unit fractions. Other students did need the model to solve. Their deep understanding of fractions as iterations of unit fractions helped them to use either a tape diagram or a number line to come up with an invented strategy and solve. This difference is important because as a teacher, I want students to be able to do both. Models can and should help students to problem solve. I want students to have enough mathematical confidence to look at a problem they've never seen before and determine what actions they can take to be successful. Additionally, I want them to be able to reason mathematically and use models to prove that their solutions are correct. Both skills are valuable, but knowing which students are doing each also helps me as a teacher to know where they are at in terms of mastery.

Using questioning to construct viable arguments and critique: There is a difference between explaining your work with evidence versus creating a viable argument to see if your reasoning is correct. As teachers, we need to come up with questions to give students opportunities to create viable arguments. In this lesson, students were successful in modeling with mathematics and explaining their models to other students. Yet, there was a lack of disagreement, and therefore mathematical arguments did not happen authentically. As teachers, we have the power to create space for students to critique others' solutions and debate about math. For example, showing the solution 5/20 L at the end of the lesson would allow students the chance to disprove that hypothesis by using their mathematical models and evidence-based reasoning. It pushed students to not only explain and defend their own solution but to also disprove an answer by using the models and reasoning that they had already come up with. Personally, watching young mathematicians critique and debate during problem-solving discussions is my favorite part of teaching. During my planning process, I think deeply about what questions or solutions I may need to display to create that debate.

References

Common Core State Standards Initiative. (2010). *Common Core State Standards for Mathematics*. http://www.corestandards.org/Math/

Koyo Publishing. (2015). *Sansu Math*. Koyo Publishing Inc.

Student learning: division with remainders

Division with remainders, for grade 4 (9- and 10-year-old) students

Sara E. Liebert

My experience teaching division is that once problems involve two-digit divisors, where students' number sense alone is insufficient for them to reason mentally, and I try to move toward some kind of procedure, everyone's tension rises – both mine and the students. In grade 4, learning division gets especially tough for students the moment the numbers are out of their comfort level. The moment can be different for each student, but for most students division becomes especially complex with two-digit divisors and greater dividends, where reasoning mentally isn't sufficient and keeping track with paper and pencil is helpful and/or necessary. That's when the standard algorithm comes in and can be useful. But once I shift into introducing an algorithm for the sake of efficiency, certain difficulties usually arise. Somehow, the class's focus on exploring and understanding division shifts to a focus on learning how to do a particular procedure, and students' different levels of numerical comforts come into play. I've tried all sorts of methods, and never have felt completely successful. Estimating is important, but getting accurate answers can be painful.

We know that students learn new ideas by connecting them to what they already know. Students learn about multiplication before they are introduced to division, so it made sense when designing a unit on division to build on their existing experience and understanding. In the same way that we want students to understand the inverse relationship between addition and subtraction, and use that information when calculating mentally, we wanted the same for students with multiplication and division. Our team kept this in mind as we built this grade-4 unit on division with remainders for year-4 students.

At the time that this unit and public lesson were developed, our school was in its second year of whole-school Lesson Study. As a staff, we developed a research theme together: *nurture students' mathematical agency and identity through the design of lessons that engage students in a Teaching Math through Problem Solving (TTP) approach and the use of productive conversations.* Furthermore, our school developed a Theory of Action that was at the center of our unit and lesson planning. As a school we believe that *if teachers apply a TTP, then students will deepen their conceptual understanding of mathematics. Increasing students' conceptual understanding will help to support procedural fluency in math. This will result in students beginning to see mathematics as accessible through effort and*

DOI: 10.4324/9781003230915-21

identify themselves as powerful math thinkers. Students are able to communicate their mathematical ideas, and revise and reflect on them in classroom discussion and journals. Our theory of action and research theme help us as a school and as individual educators to hone our ability to notice and understand student learning.

The same year that we developed this unit and lesson, our team of grade-4 and grade-5 teachers were making a shift in how we teach mathematics by using the TTP approach. This approach allowed us to base our instruction on students' own ideas and to move the units forward in response to their understandings and misconceptions. This teaching approach gives us the opportunity to spend more time in the beginning of the unit on models that are more concrete, and to introduce more representational or abstract models as students gain more understanding. Specifically, when designing our division unit, we decided to spend more time on the use of manipulatives and equal-group drawings in order for our students to develop robust understanding of division using concrete representations. We believe that once students begin to understand partitive and quotative division and these different situations, we then can support students to move to tape diagrams and complex equations, for example. During this unit, it was also important to us that we will support students to be able to think of different ways they can interpret the remainder, regardless of the model they use. When we speak of interpreting the remainder, we are considering the following three interpretations for grade-4 students: (1) the remainder can be discarded, leaving a smaller whole number answer; (2) the remainder can force the answer to the next higher whole number; (3) the answer can be rounded to the nearest whole number for an approximate answer (Van de Walle, Karp, & Bay-Williams, 2010).

The purpose of this unit is to develop students' conceptual understanding of division, specifically their conceptual understanding of what the remainder of a division problem represents and to build students' ability to easily interpret how the remainder relates to the unit builds on what students learned in third grade, when representing multiplication and division problems and building fluency of multiplication and division within 100.

Within the public lesson plan,[1] the goal is for students to further their understanding of division with remainders by encountering a new situation – cases in which the quotient $+ 1 =$ the answer. Students analyze the question in order to recognize the situation in the story as being different from situations with remainders they have experienced. Students build on work in previous multiplication and division lessons towards understanding the connection between their equations and the problem situation, including understanding what each number in an equation represents, and understanding the interpretation of the remainder in this new problem situation.

We designed this lesson to be taught using a TTP approach. TTP deepens students' conceptual understanding of mathematics, because students communicate and revise their mathematical ideas both in their math journals and through partner talk and whole group discussions with classmates. We believe that these experiences, supported by carefully designed lessons, will nurture students who see mathematics as accessible through effort and who identify themselves as powerful math thinkers. The lesson plan given next moves from more abstract solutions to more concrete visual representations – the reverse of how solutions are usually presented. This choice was made because many of the students were relying too heavily on concrete representations of division, and we wanted to push them towards seeing the conceptual relationship between multiplication, division and the concrete representations of them.

Through careful observation during the public lesson, as well as through the post-lesson discussion and the expert feedback, our team had two takeaways that our whole school then chose to implement school-wide. First, we learned that we can create problem-solving tasks that students are motivated to solve on their own. Motivation can come from both the problem itself and arc of the lesson. Of equal importance, we learned that problem-solving tasks are powerful when they intrigue students and are connected to students' lives. By nurturing students' mathematical agency and identity through the design of lessons, we are able to engage students in problem solving and productive talk.

Lesson plan

Title of the lesson: Interpreting the Remainder in Division
Students' school: John Muir Elementary School, San Francisco, California
Student ages: 9–10 years
Instructor: Sara E. Liebert
Co-authors: Rashida Carter and Joe Mannarino
Date: December 14, 2017
Goal of the lesson: Students will further their understanding of division with remainders by encountering a new situation: cases in which the quotient + 1 = the answer. Students will analyze the question in order to recognize the situation in the story as being different from situations with remainders they have experienced. Students build on work in previous multiplication and division lessons towards understanding the connection between their equations and the problem situation, including understanding what each number in an equation represents, and understanding the interpretation of the remainder in this new problem situation.
Learning standard: CCSS.MATH.CONTENT.4.NBT.A.2: Read and write multi-digit whole numbers using base-ten numerals, number names, and expanded form. Compare two multi-digit numbers based on meanings of the digits in each place, using >, =, and < symbols to record the results of comparisons (Common Core State Standards Initiative, 2010).

Lesson flow

Posing the problem

How many of you have been on roller coasters before? How many of you like going on roller coasters? Today we have a new problem to think about. I want you to imagine that the class is going on a field trip to Great America.

There are 27 children. They are going to ride a roller coaster. Each car holds up to 6 passengers. "Let's think about how to solve 27/6."

Anticipated responses

(1) Direct modeling with manipulatives: Some students may use manipulatives to figure this out. Students also may not use manipulative but will model this in their notebooks.

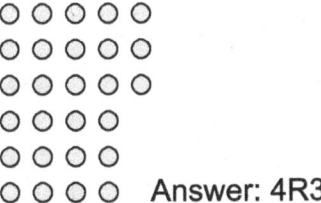

Figure 4.11 Some students use manipulatives

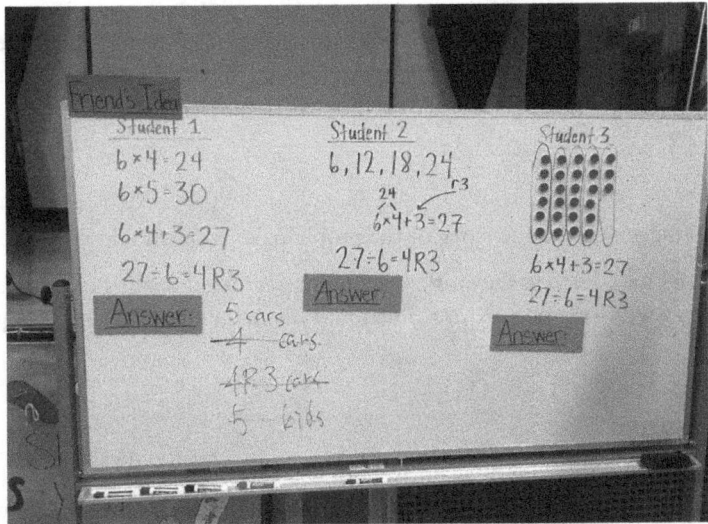

Figure 4.12 Board work plan

(2)   Use repeated addition: $6+6+6+6=24$ with 5 kids remaining. Answer: 4 cars
(3)   Use their understanding of multiplication and do $6 \times 4 = 24$ and $24 + 3 = 27$, $6 \times 5 = 30$ is too many. So, 5 is left, and the answer is 4R3. Answer: 5 cars
(4)   Skip-counting: Students may use their understanding of skip counting and do 6, 12, 18, 24, 30. Answer: 4 or 5 cars

Comparing and discussing

Board work #1: multiplication

Uses the multiplication sentence to answer the question:

$6 \times 4 = 24$
$6 \times 5 = 30$

Then the student realizes the 30 is too big, so $6 \times 4 + 3 = 27$. Therefore, $27 \div 6 = 4R3$

Board work #2 skip counting

6, 12, 18, 24 R3
$6 \times 4 + 3 = 27$

Therefore, $27 \div 6 = 4R3$

Board work #3: concrete model

Some students may use manipulatives to figure this out. (Counters will be available on the tables.) Students also may model this in their notebooks without using manipulatives. Answer: 4R3 Math Sentence: $27 \div 6 = 4R3$ (See Figure 4.11.)

Summing up

4R3 is not the answer to the story of the problem, so what should we do?

We should look back at the story. Depending on the context of the story, sometimes the answer to a problem in division with remainders can be the quotient plus 1.

What I learned

This public lesson was taught as our school was developing an understanding for what it means to collaborate as practitioners through Lesson Study. We were in our second year of whole-school Lesson Study – and not yet at the stage where the entire school could observe a public lesson together. Rather, Lesson Study teams of teachers observed lessons together. This particular lesson was a district-level public lesson, meaning the lesson was observed by over 100 participants. All participants were invited to take observational notes; but the planning team was the team that reflected publicly and spoke about their learning post-lesson, while the rest of the observers listened. I'm going to speak about my learnings in reference to collaboration with colleagues and also with reference to the students' learning of mathematics.

First, with adult collaboration in mind: I learned that this kind of collaboration helped me to have a deeper understanding of mathematics in vertical alignment, and understanding the math helped me think about the numbers we were asking in the problem, as the problem changed many times. It helped me to think about the guiding question and the new learning for the lesson. It helped me realize why the numbers we ask students matter so much. Finally, it helped me to see and hear about what other classrooms are doing, what the students are doing and what students outside of my classroom are coming up with, which allows me to further reflect on my own classroom teaching and facilitation of student learning.

With student learning in mind, I learned that we are able to create tasks that motivate students to solve them on their own and that that motivation can come in the question or the arc of the lesson. We learned that problem-solving tasks need to be connected to students' lives – it should intrigue the students. The task should be something they are motivated on their own to solve. That motivation can come in the question itself or in the arch of the lesson. By nurturing students' mathematical agency and identity through the design of lessons, we are able to engage students in problem solving and productive talk. Our team also learned that it is at times valuable, when teaching mathematics, to start with the presentation of students' more abstract solutions and move back to more concrete thinking. The students in my class for this particular lesson were very comfortable using concrete models to represent their mathematical ideas about division with remainder. So it was time to begin to move them toward larger numbers, where concrete representations were no longer going to be efficient and useful when understanding the algorithm for division. By designing a board plan (and discussion) that highlights the abstract first, we were able to facilitate student understanding of abstract concepts within division that ultimately prepared them to move toward the understanding of the division algorithm with larger numbers.

I also learned that by teaching through problem solving (TTP) in the mathematics classroom, we are able to make the space in our instruction for students to use strategies they are comfortable with. As we continue to see students' different strategies shared through partner talk, whole-class dialogue, and board work, we push students to adopt new strategies as they are introduced by classmates and as they make sense to the students. I've learned that with this approach to teaching, along with the careful design of units, we were able to build students' conceptual understanding of division as well as students' math agency and identity.

Impacts on my own work

Our team made a strategic decision to go from abstract to concrete during this lesson. We believed that this would support student thinking and their ability to understand the connection between the abstract (algorithm) and concrete representation such as arrays. We spent a great deal of time researching, discussing and finalizing the plan to move from abstract to concrete as it is the opposite of what many teachers learn is the progression that will best facilitate student learning. As a result, this has had a large impact on how I develop and create units and lessons; and now as a school administrator I ask my teachers to develop their observational tools, lesson plans and board work. We decided as a team that as a result of our learning from this public lesson, as a whole school we needed to commit to the following:

- Teachers should seek what their students' most used strategy is and then represent that in the board work.
- While we need to pay attention to language and mathematical concepts, we should also focus on overall general intellectual concepts, such as going from abstract to concrete when planning and designing units.
- Throughout the school, each Lesson Study team should think about the models they will use to represent mathematics and the relationship of this to the final mathematical expression.
- As a whole school, we need to think more deeply about the models we are using from grade to grade to represent mathematical concepts.

Note

1 The full lesson plan and edited video of the public research lesson and pre-lesson discussion can be found at https://lessonresearch.net/resources/schoolwide-lesson-study/go-public/.

References

Common Core State Standards Initiative. (2010). *Common Core State Standards for Mathematics*. www.corestandards.org/Math/

Van de Walle, J., Karp, K., & Bay-Williams, J. (2010). *Elementary and Middle School Mathematics: Teaching developmentally*. Allyn & Bacon, Pearson Education.

Student learning: multi-digit multiplication algorithm

Multiplying a two-digit number by another two-digit number, for grade 4 (9-year-old) students

Berenice Heinlein

Teaching the standard algorithm for multi-digit multiplication was challenging for me every year. When I first began teaching, this lesson felt tedious, repetitive, and unsuccessful. At the time, most of my fourth graders often forgot vital procedural steps to the algorithm because they didn't understand why the multi-digit multiplication algorithm was useful, or what each step of the procedure represented. It appeared to be a meaningless and arbitrary series of steps that they had to memorize and execute. Once students showed success with the algorithm, it wasn't due to higher understanding, but rather due to weeks of practice. This mirrored my own experience with learning to multiply multi-digit numbers in primary school.

As I improved as an educator and better understood multiplication, I was able to create lessons in which students used their understanding of place value and single-digit multiplication to reason through multi-digit multiplication. Students used a variety of strategies and models to express their thinking, and to show what they understood about multi-digit multiplication. Students used partial products and area models to decompose multi-digit numbers to multiply, depending on the context. I felt much more confident about teaching multi-digit multiplication, and students were able to calculate correctly using various strategies, and were able to create viable arguments for their decision making. The unit would go well, until it was time for me to teach students the standard algorithm for multiplying a two-digit number by a two-digit number. I struggled to help students connect this new notation to what they already knew. Many students saw the algorithm as a completely new idea. Students struggled to articulate how the algorithm was in any way similar to the strategies they employed. By the end of the unit, they understood multi-digit multiplication using models, and could work through the procedural steps to the algorithm. Nonetheless, I was concerned that the disconnect would lead to problems for them in the future, especially as they learn to multiply decimals.

By going through the Lesson Study process, I learned how to present a problem and facilitate a discussion that helps students see the connection between informal calculation strategies and the standard algorithm. Through research, I discovered strategic number choices that help students track the standard algorithm process more easily. Lesson Study was a new endeavor for our school, and Teaching

DOI: 10.4324/9781003230915-22

through Problem-solving (TTP) was fairly new to the team, but we wanted to learn more about ways to support student learning. By engaging in this process, I saw that TTP provides students the opportunity to authentically reason through mathematics.

The team met and came to an agreement on how we wanted to teach the multi-digit-by multi-digit multiplication algorithm. We noticed similar trends in our students' reactions toward learning algorithms and decided that teaching algorithms conceptually, rather than purely procedurally, was a shared area of interest for us. Although research wasn't typical practice in our school's professional development, our Principal supported Lesson Study and encouraged us to use it as a format for our learning. The team had varying levels of experience with the process, so we received support from an organization, the Lesson Study Alliance.

This research lesson was written to introduce the algorithm for two digits multiplied by two digits, taught to a class of 9- and 10-year-old students in a fourth grade classroom. This statement is the result from our initial conversations, and became a foundation for the research lesson.

> **We chose this topic because teaching the multiplication standard algorithm is consistently a conceptually difficult topic. We would like to push away from teaching the algorithm as a step-by-step procedure, and instead provide an opportunity for students to use prior knowledge of the multi-digit by one-digit algorithm and previous multiplication strategies to make sense of the two-digit by two-digit algorithm.**

This research lesson occurred at the start of the unit to examine the transition from multiplying a two-digit number by a multiple of ten, to multiplying by two-digit numbers that were not multiples of ten. The lesson focused on a school research theme:

> We want to help our students approach mathematics as a process of inquiry, rather than a process of answer-getting. . . . We hope to achieve this goal by improving our practice of teaching mathematics through problem solving, with a focus on student discussion. Student discussion will be fostered by strategic and consistent notebook use, frequent opportunities to justify and critique mathematical strategies both in small and large groups.

Lesson plan

Title of the lesson: Two-digit by two-digit multiplication
Students' school: Helen C. Peirce School of International Studies, Chicago, Illinois
Student ages: 9–10 years
Instructor: Berenice Heinlein
Co-authors: Evan Trad, Suzanne Schaefer

Date: October 26, 2018

Goal of the lesson: Students will understand that when multiplying two-digit by two-digit numbers using the standard algorithm, they first multiply the multiplicand by the ones place of the multiplier, then multiply the multiplicand by the tens place of the multiplier, and then add those products. They connect these procedural steps to what they have previously learned about multiplying two-digit by one-digit numbers and two-digit numbers by a multiple of ten.

Learning standard: CCSS Math 4.NBT.5: Multiply a whole number of up to four digits by a one-digit whole number, and multiply two two-digit numbers, using strategies based on place value and the properties of operations. Illustrate and explain the calculation by using equations, rectangular arrays, and/or area models (Common Core State Standards Initiative, 2010).

Lesson flow

Introduction

Today's problem is then shown with a blank for the multiplier:

"Ms. Schaefer has ___ students in her class. She wants to give each student 12 pieces of paper for a stapled packet. How many pieces of paper will she need to use?"

Teacher puts 20 in the blank and asks students to generate the number sentence and discuss how to solve it.

Posing the problem

Teacher now switches the number to 23.

"Ms. Schaefer has 23 students in her class. She wants to give each student 12 pieces of paper for a stapled packet. How many pieces of paper will she need to use?"

Class decides on a number sentence and discusses the difference between this question and the one that came before. Teacher helps the class formulate a guiding question for the day by altering the guiding question from the day before. "How do we multiply two 2-digit numbers, now that the ones place has a number that is not zero?"

Anticipated responses

R1: In a variety of forms, the student understands:

$12 \times 3 = 36$
$12 \times 20 = 240$
$240 + 36 = 276$

For example: Student uses an area model with a 2x1 grid or partial product strategy leaving the multiplicand (12) as a single unit and splitting the multiplier into tens and ones, ending with two products and adding them together. Or with an algorithm, student multiplies 12×20 and 12×3 and adds the products together to solve the problem.

R2: Student uses an area model with a 2x2 grid or partial product strategy splitting both the multiplicand and the multiplier into tens and ones, ending with four products and adding them together.

$2 \times 3 = 6$
$10 \times 3 = 30$
$2 \times 20 = 40$
$20 \times 10 = 200$

R3: Student multiplies 2×3, and 10×20 to get the product of 206.

Comparing and discussing

Questions for R1:

- Does this response make sense?
- Who else did something similar?
- For the 23 and the 12 – which number represents the papers, and which number represents the students?
- Where does the 240 come from? Where does the 36 come from?
- Why do you think the student split the 23 into 20 and 3, but kept the 12 as one unit?

Questions to connect R2 to R1:

- What were the four calculations that the student did to solve?
- Based on these calculations, 2 papers get passed to 3 students. Then 10 papers get passed to 3 students. Then . . . Can you finish the story?
- How do these four calculations make sense in the context of the story?
- How is this method similar to the first response? How is it different?

Questions to connect R3 to R1:

- What were the two calculations that the student did to solve?
- Where does the 6 come from? Where does the 200 come from?
- How do these two calculations make sense in the context of the story?
- What might a student who answered in this way have been thinking while solving?

Summing up

As a summary, show the standard algorithm broken into the three steps of the procedure. Lead students through the procedural steps of the standard algorithm, connecting each step back to the student methods. All students write the steps into their notebooks.

Teacher asks students, "What might be a good summary for what we discovered today?" Two or three share, and the teacher combines ideas into a statement.

Board plan

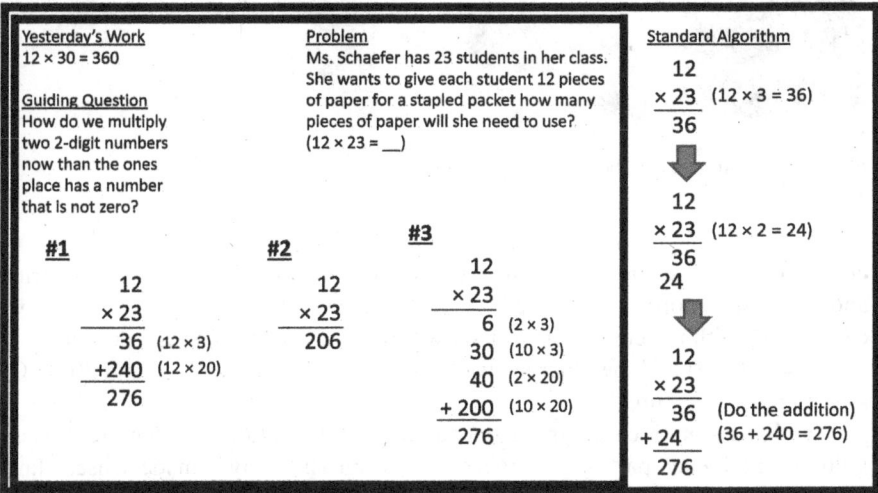

Figure 4.13 Recreation of the board plan

Observations from the lesson

The lesson began with students recalling the learning from the previous day, discussing the problem and the essential question. The teacher shared the context of the problem, leaving the multiplier blank. The teacher put 20 in the blank, asking students to generate the number sentence and discuss how to solve it. Students expressed that the number sentence is "12 × 20" quickly and confidently, and were able to express how they would solve the problem, making a connection to the previous day's problem. The teacher then changed the multiplier from 20 to 23 and asked students to consider how the problem changed. Students recognized the difference, and the teacher edited the essential question from the day before to now state, "How do we multiply two two-digit numbers when the ones place has a value that is **not** zero?" Students began to solve 12 × 23 independently, many employing a partial products method written vertically. Most utilized 12 × 20 from the introduction and extended the strategy to include 12 × 3. Some students

 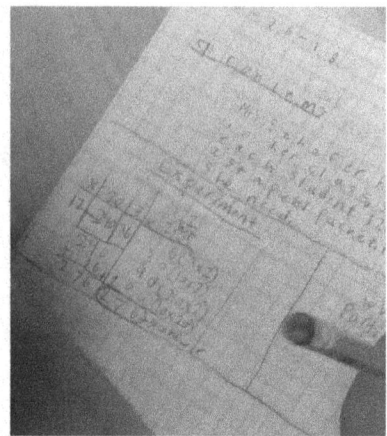

Figure 4.14 Student notebooks during independent work

chose to use an area model to visualize the multiplication. During the comparing and discussing portion of the lesson, two representative strategies were showcased, both with the correct answer shown. One showed two partial products (12 × 20 and 12 × 3) and the other showed four (3 × 2, 3 × 10, 20 × 2, and 20 × 10), as seen next on Figure 4.14.

The discussion focused on understanding both featured student responses, going through each part of the process and discussing why it made sense. Many students struggled to understand why Julia got four partial products in this lesson. The context was also a focus of questioning (e.g., "What does this 2 represent?" "2 papers from the packet") throughout the lesson. After students understood how those strategies were connected, the teacher showed the class a poster with the standard algorithm. The teacher led them through the procedural steps of the standard algorithm, and asked them to make connections between each step of the algorithm and steps from the showcased student methods. Students made important connections, but time ran out before they could fully reason through the algorithm. All students wrote the steps into their notebooks and then wrote a reflection on their learning.

In the post-lesson discussion, participants considered the pacing of the lesson and its effects on student learning. Some discussion focused on the pacing of the lesson within the unit. Instead of using various strategies for students to develop their understanding of multi-digit by multi-digit multiplication before reasoning through the standard algorithm, this lesson asked students to make sense of the standard algorithm first. The lesson didn't dedicate enough time to the algorithm and mainly focused on understanding other strategies. A secondary point of conversation focused on pacing within the lesson. As the closing speaker said during final comments, "The lesson really has two parts: 1) carrying out the multiplication of 12 x 23, and 2) making sense of the steps of the vertical algorithm now that we are multiplying two 2-digit numbers." The secondary part was only addressed in the summary, so students didn't have enough dedicated time to truly explore the standard algorithm.

What I learned

Through this Lesson Study, I more deeply understood what students need to learn from the multi-digit multiplication algorithm, as well as ways to support that learning. During the research phase of planning, I learned that it is important for students to recognize that although the standard algorithm looks different, the fundamental ideas of calculating numbers remain the same as other strategies they already know. I learned that it is more advantageous and powerful for students to break apart the multiplier (23) into place values and keep the multiplicand (12) as a whole because it more clearly represents the algorithm. Setting up the numbers in this way deliberately provoked this line of thinking.

The introductory sections of the lesson clearly connected the students' previous understanding to the lesson's new problem, which was part of the goal. By participating in the design of this research lesson, I recognized that this was a critical idea that I overlooked previously. In the rationale for lesson design, we determined that

> IF we begin the lesson by asking students to first consider the calculation 12×20, THEN students will use what they know about 12×20 to solve 12×23. For example, they might decompose 23 into 20 and 3 as they work the problem. They might also refer to 12×20 when discussing their methods.

Figure 4.15 shows that the student multiplied both two and four partial products, using two different visualizations. In the reflection, this student wrote about how the algorithm is similar and different to their original approaches. In the following lesson, when given dedicated time to reason through the standard algorithm, students shared that this lesson supported their understanding of how the algorithm can work. For example, one student said, "Oh, I see now! When doing the algorithm, you think like Julia's brain [4 partial products] but you write like Wade's [2 partial products]." I learned that students were better able to connect prior learning with new learning when it is an explicit part of lesson design. This key idea guided student thinking and helped students understand the standard algorithm for multiplication.

Impacts on my own teaching

Teaching the standard algorithm for multi-digit by multi-digit multiplication was an area of concern for me every year. My students didn't see the connection between their informal approaches and the standard algorithm. By engaging in Lesson Study, I was able to develop an idea for overcoming this challenge

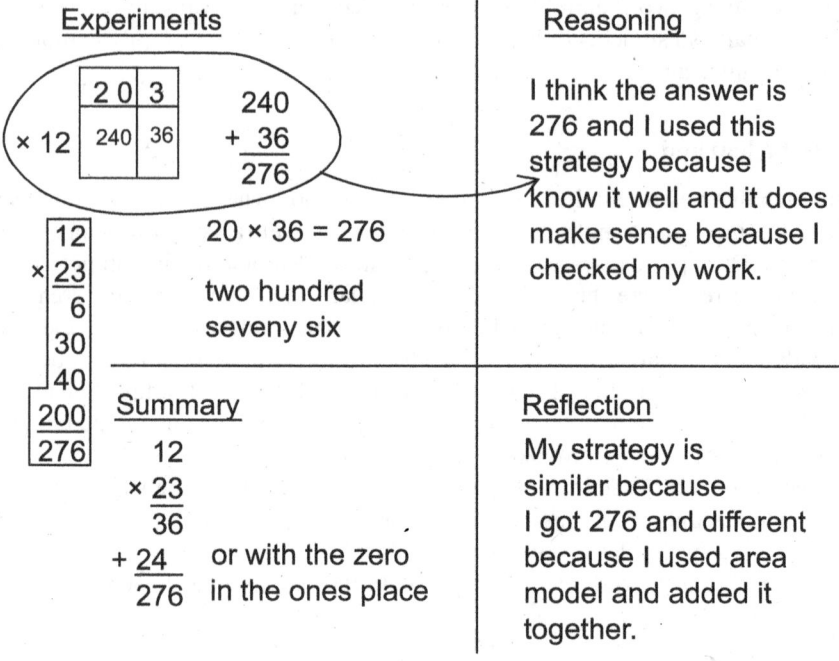

Experiments	Reasoning

Reasoning

I think the answer is 276 and I used this strategy because I know it well and it does make sence because I checked my work.

20 × 36 = 276

two hundred seveny six

Summary

12
× 23
36
+ 24   or with the zero
276   in the ones place

Reflection

My strategy is similar because I got 276 and different because I used area model and added it together.

Figure 4.15 A student notebook during independent work with a summary and reflection

by carefully designing a lesson with my team. My students better understood the standard algorithm and how it aligns with what they already learned about multiplication. This experience shaped the way that I approach multi-digit multiplication in my class. After years of me modifying my approach to teaching this content, student learning was most impacted after I experienced investigating the cause of the challenge. Now when I see a challenge in my teaching, I choose to work with colleagues using Lesson Study.

Reference

Common Core State Standards Initiative. (2010). *Common Core State Standards for Mathematics*. www.corestandards.org/Math/

Summary of Chapter IV

What we learned about student learning

Akihiko Takahashi

The Japanese professional development practice Lesson Study was introduced outside Japan to help teachers improve their lesson plans and pedagogy. Such is the goal of most professional development, but teachers who engage in Lesson Study say that it goes beyond the typical professional development they have experienced. In a typical professional development program, participating teachers expect to receive new knowledge and innovative ideas from the experts of the field. In Lesson Study, however, participants develop their own new ideas for how to best support their students to learn better. They do this through collaboration: by planning, observing, and reflecting on each other's research lessons. Thus, Lesson Study can fairly be called teacher research.

Lesson Study typically begins by identifying the difficulties and challenges teachers have observed among their students. Once these challenges are identified, the team carefully studies their curriculum standards and textbook resources to clarify what their students are expected to learn, including the learning progression in the previous and the following grades. Next, the teachers review related research as well as Lesson Study reports published by other schools to help them understand the ideas around the issues they will address through their Lesson Study cycle. This process helps them clarify their thinking and approach by addressing what the students are expected to learn as well as all possible obstacles.

All the authors of Chapter 4 report that they began their Lesson Study cycles by identifying a critical challenge in student learning at their schools. For example, both Kari Laux and her team (Chapter 4.1) and Rebecca Reddicliffe and her team (Chapter 4.2) selected the same fundamental issue: using base-ten place-value notation to understand numbers and operations. Both authors observed their students struggling with how to calculate – in one case with decimal numbers, in the other with whole numbers – and identified that in order to overcome this issue, they likely needed to better understand the key concept of base-ten place-value notations. Their Lesson Study reports discuss how they selected a problem to use in the classroom that would be "just right," i.e., not too hard or too easy, one which would successfully guide students to establish a strong understanding of place-value notation and its applications.

Lindsay Singer Kalt and her team (Chapter 4.3) decided to develop new ideas for how to teach fractions, a concept they observed to be a major challenge in their

DOI: 10.4324/9781003230915-23

elementary curriculum. The team studied the curriculum carefully and realized that using models, such as number line diagrams and tape diagrams, is essential for students to understand fractions as numbers. Through their Lesson Study process, they also realized that students could use those models for several different purposes. By inviting other teachers to observe their research lesson, the team successfully shared this new knowledge with their colleagues.

Sara E. Liebert and her team (Chapter 4.4) worked to develop new ways to engage and motivate their students to solve challenging problems since their students often do not see the usefulness of mathematics. The team discussed what kind of problems would encourage their students to learn and use mathematics in their daily life. This is a critical issue for all students, particularly in inner-city schools with an underserved student population.

A common complaint from parents and teachers is, "Why does the curriculum teach so many different ways to do basic calculations when students usually just get stuck using a clumsy approach?" Berenice Heinlein and her team (Chapter 4.5) worked to address why their curriculum encourages students to come up with various approaches for multi-digit multiplications and how to best design lessons to help students see the connections among these approaches as well as the benefit of mastering algorithms and becoming fluent in basic multi-digit calculations. To answer such questions, Heinlein and her team needed to know how students come up with their calculation methods and how they unpack the mechanisms of the calculations. Through Lesson Study, they found a way to support their students to see the relationship among the various curriculum approaches and help them understand the mechanisms behind the algorithms.

All the authors of Chapter 4 describe how they began their Lesson Study cycle by identifying a specific learning issue at their school. This is one of the critical steps in Lesson Study: determine an issue to address through collaborative teacher research. Doing so helps ensure that the results found through Lesson Study can be useful to more than just the students in a specific classroom. These results can help students in other classes and other grade levels. This is why the Lesson Study team invites other teachers in the school to their research lessons. We need more research lesson opportunities in schools as a part of the regular schedule so that more teachers can participate in planning teams and more teachers can observe and discuss research lessons to build shared knowledge about how their students learn.

V What we learned about teacher collaboration and leadership

DOI: 10.4324/9781003230915-24

Teacher collaboration: the growth of Lesson Study in one school

How the development of Lesson Study can be seen through decimals

Cassie Kornblau

On January 28, 2020, a team of six educators sat in a fourth-grade classroom after school, putting the final touches on our research lesson that was to be presented the next day. Conversations were being held about how our board work could be clearer – how the representation of a piece of ribbon mattered if the point of the lesson was for students to determine the exact length. Teachers were shouting out different ideas of the placement of the ribbon, and various drafts of the board were being drawn out. In another corner, two team members were proofreading the research lesson document before sending it off to the printer. Our assistant principal walked in to ask if we needed anything before tomorrow. I handed her a draft of sub coverage for the middle school teachers, a schedule of tomorrow's event including the pre-lesson discussion, research lesson and post-lesson discussion and responded, "I think we are good." At almost six thirty, we stopped – discussions about how to teach math could always continue – but we felt this lesson was ready. Ms. Witt, the teacher of the research lesson, said she needed time to process and practice the new changes we had made before tomorrow. As I looked around the room, our "math team", which had been just two of us five years ago, was now composed of teachers who taught math from kindergarten through eighth grade, including a special educator. It was at this point that I realized Lesson Study was not just an idea; it was part of the culture of our school.

The evolution of Lesson Study

When I started at Brentano seven years ago, our assistant principal introduced me to Lesson Study. I was part of a team that designed an English language arts lesson about text features, and our principal taught it to a class of sixth graders because the rest of us were afraid to volunteer. I went through the process and was still a bit unsure whether I liked it or not but was interested in seeing how it could translate when dissecting a math topic. The following year, the new seventh and eighth grade math teacher and I joined forces and committed to Lesson Study to help achieve two goals: to improve our instructional math practice, and to begin to spread the idea of teaching math through problem solving (TTP) from kindergarten through eighth grade.

DOI: 10.4324/9781003230915-25

Over the next three years, teams of Brentano teachers participated in eight research lessons. The math team, composed of the three departmentalized math teachers from fourth through eighth grade and included the special educator that supported these students. Together, we conducted five research lessons including two at the Lesson Study Alliance Conference in Chicago. In addition, the kindergarten and first grade teachers designed and participated in two Lesson Study cycles, and the second and third grade teams launched three research lessons.

The biggest change that came from all math teachers participating in Lesson Study was a shift in how math was taught. Teachers moved away from modeling how to solve problems for students to following the methodology of TTP. Teachers began to approach topics conceptually whereby tasks were chosen that lend themselves to deep discussion and presentations of different students' strategies. At the same time, teachers watched the camaraderie develop among members of the "math team" and wanted to join in. They saw how close we had become as colleagues, willing to argue, discuss and push each other's thinking. As a result, in addition to the grade-level teams that annually planned research lessons, the "math team" expanded and welcomed all who wanted to join – making it a now a group that encompassed teachers from kindergarten through eighth grade.

Exploring decimals as any other number

As new research Lesson Study cycles begin, my team or the "math team" brainstorms a topic of interest based on something everyone finds difficult to teach and/ or for students to comprehend. This time around, the fourth grade teacher – who was in her third year of teaching – expressed her desire to explore decimals. She explained the curriculum she had been using did little to aid in students' conceptual understanding of decimals as numbers. Rather it was matter of fact, telling students for example, "this bag of candy is $\frac{1}{10}$ of a gram. On scale, it says 0.1 grams, so 0.1 equals $\frac{1}{10}$."

In our first few meetings as a math team, we dove into the research or kyouzai kenkyuu, exploring various curricula and their introduction of decimals. Upon exploration, we zeroed in on two big ideas that we wanted students to grasp in terms of this lesson but also in terms of how they approached numbers:

- We wanted students to see decimals as numbers in the larger picture of their base-ten place value understanding.
- We wanted to de-silo students' understanding of whole numbers, fractions and decimals as separate but rather view them all as numbers.

As a result, the team decided instead of sequencing lessons directly from teaching tenths to hundredths in fractions and then decimals, it was necessary to go back and build on the whole number place value understanding. Similarly, the team felt for students to truly understand decimals as numbers they needed to be able to place them on a number line like they had done in previous grades with whole numbers.

To design this unit, the team looked at previous work and noticed that students built an understanding of tenths and hundredths as fractions of a whole. Although students could accurately represent and compare fractions with denominators of ten and hundred, they struggled to identify and properly place the fractions on a number line. This revealed a gap in students' understanding about tenths and hundredths occupying the space between whole numbers in linear representations.

Therefore, to sequence this unit, in the first lesson, the team built on students' prior knowledge, specifically using 10 x 10 grids, partitioning numbers lines into tenths; and those tenths in hundredths, and placing the fractions more precisely on number lines. In the second lesson, students used their understanding of tenths and hundredths as fractions to extend their whole number place value knowledge beyond the decimal point to include tenths and hundredths. At this point, students had practiced representing decimals numbers including tenths and hundredths, identifying equivalent fractions to decimals and vice versa, and comparing decimal numbers by justifying their comparisons using the value of each digit.

Now, for the research lesson, the team focused on having students place tenths and hundredths linearly on a number line in decimal form because previously they had only placed fractional equivalents. The goal for this lesson was for students to understand the linear representation of how whole meters can be broken down into tenths and how tenths can then be broken down into hundredths by identifying specific lengths in decimal form on the number line. The unit would then wrap up with students using the number lines to demonstrate fractions and decimals as equivalent representations of specific points between whole numbers.

Lesson plan

Title of the lesson: Labeling Exact Lengths on a Number Line

Students' school: Brentano Math and Science Academy Chicago, Illinois

Student ages: 9–10 years

Instructor: Jessica Witt

Lesson plan developed by: Erendira Alcantara, Aaron Bingea, Blair Brodie, Cassie Kornblau, Martin Lenthe, Brittany Williams, Jessica Witt

Date: January 29, 2020

Goal of the lesson: At the end of the lesson, students will understand that decimals to the hundredths place represent specific points on the number line, created by breaking ones into tenths and tenths into hundredths. This understanding will manifest itself by students being able to identify distances marked on the number line and justify placement verbally.

Learning standards: Compare two decimals in hundredths by reasoning about their size – CCSS Math 4.NF.C.7 (Common Core State Standards Initiative, 2010)

Lesson flow

Posing the problem

I have 1 green ribbon, and I want to know exactly how long it is. I'm going to measure it using these meter sticks. This is 1 meter, this is 2 meters, this is 3 meters (mark on board).

Teacher question: "What do you notice?"

We want to know exactly how long this ribbon is, so we are going to zoom in on this section here to get more specific while we are measuring the ribbon.

On your sheet you are going to see the "zoom-in" of this section. This point shows the length of the green string.

We want to know the exact length of the ribbon. You have two minutes to independently find the exact length of the green ribbon using a decimal. Be sure to justify your thinking on your paper.

Anticipated responses

- Desired response: 1.91 m. because 1 and 9 tenths and 1 hundredth or 1 and 91 hundredths because it is 1 hundredth more than 9 tenths and 1 hundredth or 1 more than 1 and 91 hundredths. The distance between 1.9 and 2 broken into ten parts will result in parts representing 1 hundredth of a meter.
- 2.10 because after 9 tenths is 10
- 20 because 19, 20
- 2 because 1 more tenth is 2
- 2.9 because 2 is after 1
- 1.10

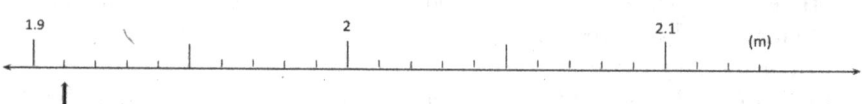

Figure 5.1 Number line showing the length of the first piece of green string

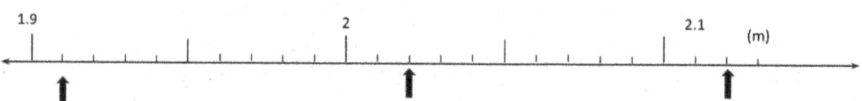

Figure 5.2 Number line showing additional lengths of the pink and black ribbon. The pink ribbon is 2.02 meters long. The black ribbon is 2.12 meters long

Posing the problem, part II

I have two more ribbons; I want to know exactly how long these ribbons are. Tape up pink ribbon and black ribbon.

You have two minutes to independently label the length of the pink ribbon and the black ribbon on your number line. Be sure to justify your thinking on your paper.

Anticipated responses: black ribbon

Desired response

- 2.12 m. because it is 2 hundredths past 2.1

Misconceptions

- 2.3 because I counted 2 tenths after 2.1
- 2.03 because I thought it was 2.3, but then I remembered from the last one we are counting by hundredths, so I changed it to 2.03

Anticipated responses: pink ribbon (2.02 m)

Desired response

- 2.02 m. because it is 2 hundredths past 2

Misconceptions

- 2.2 because it is 2 tenths after 2
- 4 because 2, 3, 4
- 2.92 because point a was .91, and it is one more space, so it is .92

Summing up

I can be more exact in my measurements between whole numbers by using tenths and even more using hundredths.

How has this lesson changed my practice and the practice of my team?

When I reflect on this lesson, I am drawn to two ideas: how this lesson improved my understanding and teams' understanding of how to teach decimals and how this lesson represents the growth of teacher learning and leadership at my school. In this lesson, we observed our students productively struggle with the placement of decimal numbers on a number line. Upon further reflection on the progression of concepts that deal with decimals and numbers in general, we realized that our students need more time with one dimension measurement and number placement – in essence, more work with numbers lines. In turn, this lesson has started a larger discussion around the consistency of number line usage in all grades and the process of norming around how we teach placement of numbers. With representatives of teachers from kindergarten and third, fourth, fifth, sixth, seventh, and eighth grade on this research lesson team, understanding why we need to use the number line now becomes even more applicable to not just this lesson but the school.

Similarly, this lesson proved students need familiarity with applying the decomposition of a number line into ten equal pieces to represent the base-ten number system using whole numbers. Therefore, precisely placing hundreds on a number line from 0–1000 by partitioning the number line into ten equal pieces needs to come before this occurs with decimals and fractions in fourth grade – a realization primary teachers on this team see the need to implement immediately. Further, this lesson brought to life the need to teach tenths and hundredths separately in terms of decimal place value and number line placement. In essence, students should master tenths before moving onto hundredths. Although Common Core has tenths and hundredths being introduced and taught in fourth grade, the team began to question and discuss whether the concept of tenths should be moved to third grade – like in Japan – to provide students with practice of decimal numbers in tenths before seeing hundredths a year later.

In addition, this lesson highlighted for my team and me how the context of a problem often determines whether students will grasp a concept. In this case, giving students a familiar context of length with a clear visual of ribbon provided an access point for all students to join into the discussion, and created excitement to find the precise lengths of the ribbons on the number line. Had the context been more abstract, I feel the students would have struggled to determine what was being measured and how the solution related back to the real-world situation.

Yet, besides the rich discussions that came out of this lesson on how we should teach decimals, its connection to fractions and whole numbers, and the usage of number lines throughout the grade levels at my school, this lesson embodies for me a cultural shift that teachers have the power to drive their own learning and lead each other through the process of Lesson Study. This cycle was entirely teacher-directed – no administrators required any teachers to be present, nor did they impose any restrictions on this research lesson. The administrators rather trusted the leadership of teachers who had done this before to lead others who were less familiar and in essence allowed us to learn together.

When I ask colleagues that I work with daily what Lesson Study at Brentano means to them, I often hear words like it is a lifeline, a place to get help, to get better, where no judgment will be imposed on anyone – a place where we get animated, we argue all about math, and the nerdiness in all of us comes out. Often, research lesson cycles are the only place where we as teachers slow down and look closely at what we are teaching, how we are teaching it, and, perhaps most importantly, why. It provides a place where new and experienced teachers come together to build valuable relationships with their peers. It is a place where we look at examples of rich discussion, inspect multiple curricula from across the globe, and see its intersection with the standards. During Lesson Study, we share ownership – regardless of who teaches the lesson – and each cycle we learn to better plan and execute deep learning to take back to our own math classrooms. Lesson Study is not about the individual teacher but rather how a team of educators can work together to improve the learning of students. I feel fortunate that this team embodies the meaning of Lesson Study and teacher learning within the walls of my own school.

Reference

Common Core State Standards Initiative. (2010). *Common Core State Standards for Mathematics*. www.corestandards.org/Math/

Teachers solving their own problems of practice

Quotative division for grade 3 (8- and 9-year-old) students

Brigid Brown

In the fall of 2017, our school was beginning our third and final year of grant-funded Lesson Study work with Mills College. More and more teachers were beginning to utilize Teaching through Problem-Solving as an instructional approach, and we had identified Operations and Algebraic Thinking as a domain in which to focus our work. In a cross-grade-level team of second and third grade teachers, we decided to dig into an area that was a persistent struggle for our students: division word problems. Through this study, my team and I identified and committed to the use of various strategies to make our teaching more impactful. Additionally, through this experience I was able to see the powerful impact of school-wide, cross-grade-level Lesson Study as the work develops over time. In our school, Lesson Study allowed teaching teams to see the need for school-wide instructional shifts and also to lay the groundwork of professional community and trust necessary for our school to successfully embrace instructional shifts in a way that promoted significant and lasting change in the service of student success.

Our students' challenges with word problems showed themselves across the grades from kindergarten through fifth (ages 5 through 11). Despite our best efforts, many of our students approached these problems seemingly at a loss for how to choose an operation to solve, often defaulting to taking the two numbers in the problem and either adding (in grades K through 2) or multiplying (grades 3 through 5). Looking at the work our students produced independently, we saw many students grasping at an incomplete response, unclear on how to fit their work back into the context of the problem. While teachers dutifully asked students to "show their work," the inconsistent and often incorrect use of diagrams and equations seemed to demonstrate a conceptual understanding that was incomplete. How could we get our students to unpack the language of the word problems effectively, thoughtfully make use of models and equations to calculate, and recontextualize their work into the situation in the problem, all without us holding their hands through each step of the process?

The lesson our team selected was a third grade lesson at the very beginning of students' study of division: the first lesson in which they'd encounter a quotative division problem. We chose this lesson hoping to answer some questions: the ones given earlier about equations and word problems, but also some bigger picture questions about our multiplication and division teaching school-wide. We had

DOI: 10.4324/9781003230915-26

been given some suggestions by our mentors about how to better set students up for success, and we were investigating them. One suggestion was that the third grade team teach multiplication and division in a sequence, as opposed to simultaneously as our curriculum outlined. The third grade team was in the middle of piloting that shift, having written and taught a multiplication unit with a Teaching through Problem-Solving approach, and we were in the planning stages of a division unit to follow. Another suggestion on the table was that we consider beginning study of multiplication in second grade, instead of waiting until third grade, when the Common Core State Standards requires it. We were open to the possibility, but wanted time to investigate the question as a team to make sure it made sense in our school context. And finally, as we neared the end of our three-year Lesson Study project, we wanted to figure out what learning we could take from the experience of working with Lesson Study and Teaching through Problem-Solving. What had we learned, and how could we extend the learning past individual lessons and teams? To what degree could we reasonably ask teachers to incorporate TTP into their math teaching, and could we possibly marry it with our existing curriculum? What amount of unit and lesson design was reasonable to ask of teachers, and what barriers would get in the way?

At this point in our Lesson Study work, our approach to teaching had changed as a team and school-wide, and this research lesson stands out to me as a turning point. The experience encompassed in this research cycle was an example of the agency and self-efficacy our teachers had developed over the course of two and a half years using Lesson Study as a forum for inquiry into solving our own problems of practice. By now, we could see that the challenges our students faced with word problems were not isolated to any of our individual classrooms. These challenges were a theme woven throughout their math study with us, from kindergarten to fifth grade. And as such, they necessitated a unified response from our entire team in order to shift our students' internalized understanding of how to make sense and persevere in solving word problems.

Lesson plan (shortened[1])

Title of the lesson: Quotative Division: How Many Kids Will Get 3 Cookies Each if There Are 12 to Start?

Students' school: Acorn Woodland Elementary, Oakland, CA

Student ages: 8–9 years

Instructor: Ginger Cook

Co-authors: John Aragon, Elena Cabañas, Brigid Brown, Jana Morse, Shelley Friedkin

Date: November 1, 2017

Goal of the lesson: Students will further their understanding of division with a new situation – quotative division. Students will analyze the question in order to recognize the situation in the story as being different from the partitive situations they have experienced. Students build on work in previous multiplication and division lessons towards understanding the connection between their equations and the problem situation, including understanding what each number in an equation represents.

Learning standard: Interpret whole-number quotients of whole numbers, e.g. interpret 56 / 8 as the number of objects in each share when 56 objects are partitioned equally into 8 shares, or as a number of shares when 56 objects are partitioned into equal shares of 8 objects each – CCSS Math 3.OA.A.2 (Common Core State Standards Initiative, 2010).

Use multiplication and division within 100 to solve word problems in situations involving equal groups, arrays, and measurement quantities, e.g., by using drawings and equations with a symbol for the unknown number to represent the problem – CCSS Math 3.OA.A.3 (Common Core State Standards Initiative, 2010).

Lesson flow

Posing the problem

Owen has 12 cookies. If he gives 3 cookies to each friend, how many friends can share the cookies?

Anticipated responses

R0: misconception:

$3 \times 4 = 12$ kids

R1: direct modeling:
Using manipulatives, student puts 3 "cookies" in a group until all 12 cookies are gone.
R2: representational:
Student draws 12 cookies and then proceeds to circle 3 cookies at a time until all are grouped.
Note: looking for someone representing the problem and labeling their work.
R2 might include:
skip counting
repeated addition
starting with 12 and subtracting 3 at a time

R3: abstract:

$12 \div 3 = \square$
$\square \times 3 = 12$

(If student doesn't write equation but solves thinking about it in terms of multiplication, the teacher can ask the class how they might represent the thinking with an equation.)

Comparison and discussion

Share strategies in order R1, R2, R3 (share R0 only if shown by more than four students).

Possible question to promote understanding of specific strategies:

- What does each number stand for?
- What does this (specific number, symbol, or part of a diagram) mean?
- What equation could we use to show this?

Possible questions to make connections across strategies:

- How is (R2) similar to (R1)? What is similar and what is different?
  - o Aim for students to see how the representational maps onto the concrete.
  - o Students may notice that one starts with a total and one starts with a part.
  - o Students can understand that a given problem can be solved flexibly. When you keep good track of what the parts of your solution represent, then you can find your correct answer.

- I heard someone say this was the same problem from Monday. What do you think [if students notice that this is $12 \div 3$ like in previous problems]?
  - o Why then are we not making three equal groups? (Help students understand the difference between "groups of 3" and "3 groups of _".)
  - o What's different from how we've been dividing until now? (Help students understand the difference between the two problems and the importance of labeling to know what each number represents.)

- Did division work when we were trying to find the number of cookies? Did it work when we were trying to find the number of people?

Summing up

Possible summary (if students are not yet ready to compare types of division):
   This is a problem we can solve with division.
   Alternate summary:
   Today we learned a new division situation. We can use division both to find the # of groups (kids) or to find the # in each group (cookies).

Board plan

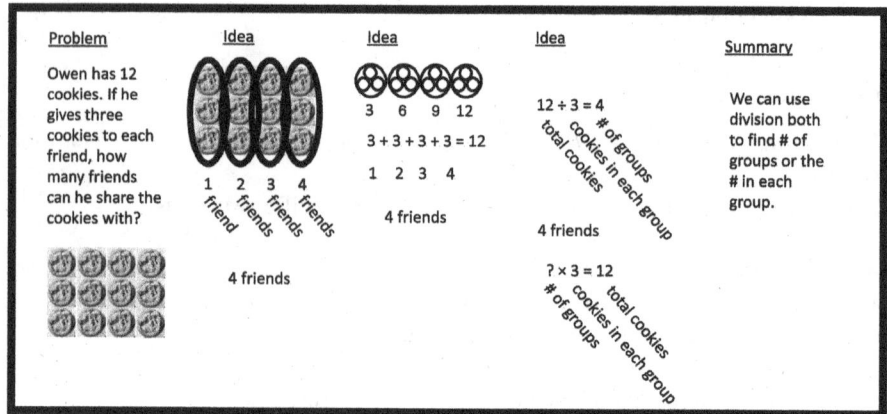

Figure 5.3 Board plan

Observations from the lesson

The post-lesson discussion for this research lesson had a notable tone of conviction and willingness to reach consensus around questions of how to teach word problems across the grade levels. Specifically, the school-wide team saw the usefulness of encouraging students to label the equations they use to represent and solve the word problems, making clear the connections between the quantities and relationships in the story problem and the numbers and symbols in the equation. At different grade levels, this might look different. A single letter or a simple drawing might serve as a label in early grades, for example. Or in the case of a striving reader or when a students' individualized education plan might specify, a verbal explanation could replace a written label.

The school-wide team was also ready to respond to the question our team posed: "What counts as an answer to a word problem?" Observers saw evidence in our research, in the students' work, and in their own experiences that suggested that a robust understanding of a word problem often went hand in hand with an answer that was expressed across multiple modalities. Students who could draw and label diagrams, explain their equations and label their unknowns, and respond with not just a number but a quantity or sentence were the same students who could explain and justify their answers. And, maybe more importantly, for students whose answers were incomplete or incorrect, prompting them to include these facets of a solution seemed to help spur their thinking, leading them to clarify their understanding or note and possibly correct an error. By the end of the discussion, teachers across all grade levels had chosen to commit to the expectation that an answer to a word problem would be complete when students (1) drew and labeled a diagram, (2) wrote and labeled an equation, and (3) wrote (or could state verbally) a sentence stating their answer in the context of the problem.

What I learned

When our planning team met the following week to reflect, we revisited some of our questions about how and when to teach certain operations across the grade levels. Through our study of the progressions of the Common Core State Standards and dialogue about our own teaching experiences, we could see that the patterns in our students' tendencies were the result of years-long experiences with the operations. Which operations and which definitions were emphasized over the course of each year gradually built up towards our students' lop-sided comfort levels: in the primary grades, between addition and subtraction, our students favor addition; in the upper grades, between multiplication and division, our students favor multiplication. We knew we had to shift our time and energy to put a greater emphasis on the operations with which our students struggled if we wanted to nurture a more balanced understanding.

We were also able to gain clarity around some questions about teaching division in third grade, based on our study of how the learning progresses across grade levels. Whereas at the beginning of our research cycle we were convinced that the third grade team needed to teach our students to differentiate and name the two types of division – quotative and partitive – we decided that in the first year of study that was not as important as having many intentional experiences building understanding of the meanings of division and multiplication and how to represent division and multiplication situations with manipulatives, diagrams, and equations. We could expect that students would shift to a more abstract understanding of division over time – so long as they were given sufficient time to experience and explore concretely early on.

Likewise, our second grade team members gained a greater clarity around their role as the last year where there's significant focus on building the conceptual understanding and procedural fluency in addition and subtraction. The second grade team felt a new urgency around solidifying students' understanding of multi-digit addition and especially subtraction, knowing that while these operations are formalized in third grade with standard algorithms, they are not a focus, and there is little time available for reteaching. The team committed to dedicating the time needed for second graders to bridge between the concrete, representational, and abstract in addition and subtraction, so that the understanding these students brought to third grade would be strong enough for them to seamlessly map the standard algorithm on top of it. This, the team decided, would contribute significantly to the work of the upper grade teachers, and the second grade team decided that would be their focus rather than introducing multiplication.

Early on in our school's Lesson Study work, there was often a question about the usefulness of spending so much time working on one lesson, particularly for teachers outside that grade level. The learning from this cycle demonstrated how impactful cross-grade level collaboration can be. By collaborating across grade levels, our team was able to situate the challenges we saw our students demonstrating within a larger arc of their learning over their elementary years. We were better able to see the learning from our students' perspective, as opposed to our own experience of teaching the same ten-month cycle over and over again. For both grade level teams, the opportunity to collaborate across grade levels in Lesson Study allowed us to gain a deeper understanding of where we fit in the multi-year progression of learning. This, in turn, led to a deeper understanding and internalization of the goals of our respective grades and a commitment to leaning into the toughest areas of instruction in order to best prepare our students for what lay ahead.

Impacts on my own work

Watching the work of the planning team unfold helped me, as an educational leader, gain an appreciation of the powerful potential of school-wide, cross-grade-level Lesson Study to shift the culture and practice of a school over time. For example, there was a noticeable difference between this cycle's research lesson day and those of some of our earlier cycles. The research lessons from our first year were fascinating, and teachers were excited to have the opportunity to observe our own students' learning in such detail. But there was a way in which each cycle felt isolated and the implications of our study somewhat ambiguous. By this point, though, there was a shift. The year before we'd made the decision to schedule our research lessons after school so that the whole staff could participate. We'd also dedicated professional development time over the summers to our math Lesson Study work, and during that time we'd revised our Research Theme and Theory of Action to be more specific and relevant. All of these shared experiences accumulated such that this particular research lesson and post-lesson discussion felt like one part of an ongoing conversation about math learning and teaching at our school: a school-wide collaborative problem-solving discussion. As a staff, we shared a growing understanding of the reasons behind our students' persistent math struggles, and we were ready to investigate ways to address them and to use our research observations and discussions as a forum for collective decision making.

Just like in a math classroom, this kind of collaborative inquiry requires both structures and community norms in order to be effective. For our school, school-wide cross-grade-level Lesson Study provided a foundational structure, upon which we gradually built other necessary structures: clear guidelines around planning and teaching practices, school-wide alignment to a standards-based curriculum, and dedicated time for collaborative planning and peer observation. Lesson Study also provided a foundation for the values, norms, and attitudes conducive to our shifts. We built an adult learning community based on professional trust and a growth mindset, where teachers could examine their own practice in collaboration, identify problems and try to solve them, and know that mistakes are a normal and necessary part of the process. Individuals were allowed multiple points of entry: whether they brought more or less experience with math teaching, or more or less interest in changing their practice, people could engage at their own pace.

Of course, there are times when school leadership must mandate instructional shifts, but there's a benefit to handing responsibility for decisions over to a larger school-wide inquiry within a Lesson Study community. It hasn't happened quickly, but the shifts our school has made have been authentic, impactful, and lasting. Lesson Study led us to dozens of small agreements around content and instructional strategies – even something as specific as what would be considered a complete response to a word problem. But we also made major shifts: a shift in curriculum and a shift in instructional pedagogy, to discussion- and inquiry-based Teaching through Problem-Solving lessons. Our students made measurable gains in their math learning, demonstrating increased competence in classwork as well as on standardized tests. And perhaps most importantly, the school culture around math changed. Teachers embraced math instruction more deeply, many saying that they enjoyed teaching math for the first time in their careers. And our students began participating eagerly in challenging classwork and naming math as their favorite subject. From my perspective, that makes it well worth the time and effort.

Note

1 A full version of the lesson plan can be accessed at https://LSAlliance.org/Lessons.

Reference

Common Core State Standards Initiative. (2010). *Common Core State Standards for Mathematics*. www.corestandards.org/Math/

Lesson Study and the new teacher

Adding fractions with unlike denominators, for grade 4 (8- and 9-year-old) students

Alexandra Johansen Laughlin

According to the CCSS-M for fourth grade, a significant area of focus for instructional time is on the development of fractions. This includes the interpretation and creation of pictorial representations, fraction decomposition, fractions as units, fractional equivalency, and adding/subtracting fractions with like denominators.

Our team's research helped us define why fractions historically have been a particularly difficult topic for elementary school students. Throughout the research process and in conversations with colleagues, fractions were deemed inherently challenging because of the myriad ways that they can be interpreted and represented. However, I have found that the confusion does not come from the concept itself, but rather from our system's clunky treatment and introduction of fractions for our students. Often, students are first exposed to fractions using circular, square, rectangular, and linear models. (Circular models are particularly problematic in the early phases of fraction introduction because the decomposition of circles to create equivalent fractions often does not result in equal parts; further, a circular representation does not readily translate into other models, like the number line.) Additionally, in some introductions of fractions, students are asked to interpret part–whole relationships and posit about proportions, e.g. $\frac{2}{3}$ of 12 is 8. This "over-complication" of the concept muddies the essence of what students need to understand during their first experiences with this new dimension of mathematics; i.e. fractions are units. This realization meant that our lesson and unit would treat fractions as countable units (1 fifth, 2 fifths, 3 fifths, etc.), not unlike other units in our world (1 apple, 2 apples, 3 apples, etc.).

In addition to seeking to demonstrate how an alternative method to introducing fractions might help students understand the concept, our research team wanted to address a gap/misalignment in the CCSS-M (Common Core State Standards Initiative, 2010) between grades 4 and 5. We believed that although CCSS-M categorizes the addition of fractions with unlike denominators as a fifth grade standard, fourth grade students are fully capable of extending their understanding of fraction equivalency to the addition of fractions with unlike denominators. Otherwise, what's the point of creating equivalent fractions, anyway? Why not allow students the opportunity to see the creation of equivalent fractions as a manipulation tool for solving more complex fraction problems?

DOI: 10.4324/9781003230915-27

Some broad mathematical goals that we had for our students included:

- deriving a mathematical expression from a story problem
- drawing accurate pictorial representations of fractions
- making connections between various representations of equivalent fractions
- articulating how equivalent fractions is a tool for solving many types of fraction problems

Some fraction-specific questions included addressing students' understanding of the following:

- what is a unit?
- how can units be counted?
- how can we illustrate and determine the size of a unit?
- how can decomposing units help solve problems?
- how can we combine like and unlike units?

More than addressing specific mathematical challenges, however, we wanted our students to have the mindsets, dispositions, and habits that would encourage them to engage with novel concepts with energized critical thinking and their own creative problem-solving strategies. In my classroom, one of my priorities has always been creating opportunities to help my students feel a sense of empowerment and ownership over their thinking.

Some habit-specific goals included:

- listen and think critically about classmates' thinking
- build off of each others' ideas
- ask and answer questions about own and others' reasoning
- respectfully disagree with classmates' ideas and explain why
- be able to both articulate own ideas and change own perspective when new credible information is presented
- celebrate own efforts and that of classmates
- make conceptual connections between classmates' ideas within a lesson and across lessons

We believed that by setting strong classroom culture and an environment of respect and rapport from the beginning of the school year, students would be comfortable and excited to participate in all subjects with a sense of agency over their learning experience. Further, by the time we asked our students to engage with fractions toward the end of the school year, they had internalized the positive mindsets and habits given previously.

From the research and planning process, my learning can be categorized into three buckets: mathematical, pedagogical, and professional. Mathematically, I learned that teachers should create opportunities for students to see connections between mathematics and life. That is, fractions are countable units, the same way we count pencils or apples. In addition, we need to help students develop a strong foundational understanding (primarily through pictorial representations) of fractional units in order to deeply understand why something like fourths and eighths are different types of units (apples vs. oranges) and thus cannot be combined until they are converted to the same unit (pieces of fruit). Through research, I learned that the concept of adding like units starts as soon as students combine numbers in Kindergarten, and this principle is woven throughout mathematics into high school, when students are asked to simplify algebraic expressions by combining like terms. Pedagogically, I reaffirmed my understanding that students thrive in the reliability of a lesson structure and communication norms that support various ways of thinking and creating. Professionally, I learned that teachers (read: myself), like students, rise to expectations when given strategic supports and encouragement; and when they feel that the work itself matters. One of the most poignant lessons for me was that with a strong classroom culture, students feel safe and excited to think creatively and to critically challenge the reasoning of their peers – two life skills that many of us adults are still developing.

Lesson plan (shortened[1])

Title of the lesson: Let's think about how to calculate $\frac{1}{4} + \frac{3}{8}$
Students' school: Dr. Jorge Prieto Math and Science Academy, Chicago, Illinois
Student ages: 8–9 years
Instructor: Alexandra Johansen Laughlin
Co-authors: Alexandra Johansen Laughlin, Mary Espinosa, Yael Berenson, Maria Rosario, Andrea Calhoun, Andrew Friesema, Mariel Laureano
Date: May 5, 2016
Goals of the lesson:

   a) Students will understand that in order to add or subtract fractions with unlike denominators they have to change one of the fractions into an equivalent fraction with a denominator the same as the other addend.
   b) Students will be able to use visual models and number sentences to find equivalencies in order to add fractions with unlike denominators. Students will be able to communicate their ideas to peers using the models. Students will be able to summarize how to convert one fraction to an equivalent fraction and why this must be done before adding fractions with unlike denominators.

Learning standards:

- Explain why a fraction a/b is equivalent to a fraction $(n \times a)/(n \times b)$ by using visual fraction models, with attention to how the number and size of the parts differ even though the two fractions themselves are the same size. Use this principle to recognize and generate equivalent fractions. CCSS Math 4.NF.1 (Common Core State Standards Initiative, 2010).

- Add and subtract fractions with unlike denominators (including mixed numbers) by replacing given fractions with equivalent fractions in such a way as to produce an equivalent sum or difference of fractions with like denominators. For example, $2/3 + 5/4 = 8/12 + 15/12 = 23/12$. [In general, $a/b + c/d = (ad + bc)/bd$. CCSS Math 5.NF.1 (Common Core State Standards Initiative, 2010).]

Lesson flow

Introduction

Teacher will begin lesson by asking class what $1/8 + 3/8$ is in order to compare and contrast this problem with the new problem for today, $1/4 + 3/8$.

Posing the problem

Hatsumon: Alonso ran $1/4$ of a mile on Monday and $3/8$ of a mile on Tuesday; how far did he run altogether?

   Potential T support: Think of a number sentence that represents how far Alonso ran altogether, and write it in your notebook.

Anticipated responses

   R1 (correct):

$$\frac{1}{4} = \frac{1 \times 2}{4 \times 2} = \frac{2}{8}$$

$$\frac{2}{8} + \frac{3}{8} = \frac{5}{8}$$

Figure 5.4 Anticipated response R1

   R1: Student recognizes that they cannot add fourths and eighths. They recognize that four is a factor of eight and that they can make an equivalent fraction for $1/4$ that has a denominator of 8 so that they can add it to $3/8$.

R2 (correct):

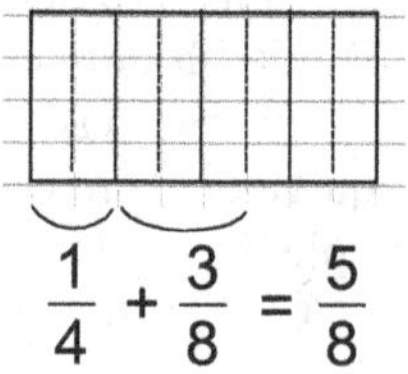

Figure 5.5 Anticipated response R2

R2: Student represents ¼ with a tape diagram and represents ⅜ with a tape diagram. Student recognizes that they can break their fourths into eighths and count up the number of eighths to find ⅝.

R3 (correct):

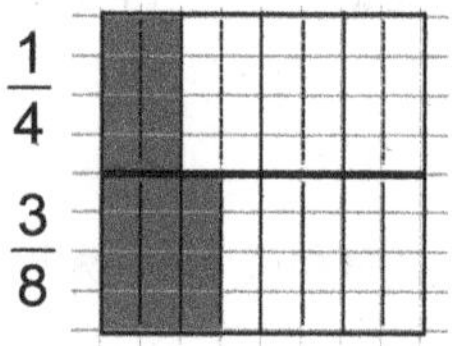

Figure 5.6 Anticipated response R3

R3: Student represents ¼ with a tape diagram and represents ⅜ with a tape diagram and does not know how to add the ¼ and ⅜.

R4 (correct):

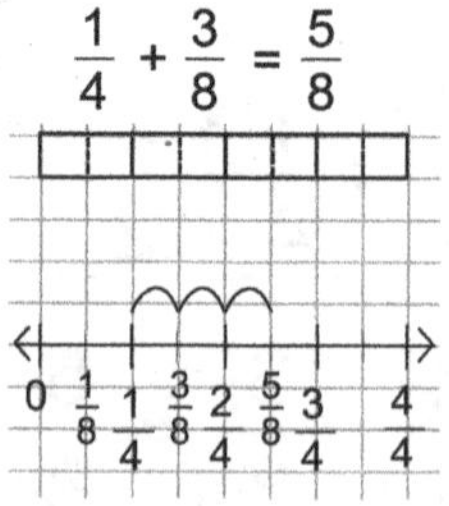

Figure 5.7 Anticipated response R4

R4: Student uses number line divided into fourths and eighths, starts at ¼, moves 3 eighths and finds ⅝ on the number line.

Note: another potential response for some groups of students may be that they add the two denominators together, e.g. fourths plus eighths equals twelfths. However, the instructor, Alex Johansen Laughlin, did not believe her students would respond in this way to the problem, given the emphasis on pictorial representations of fractions leading up to this lesson. i.e. this particular group of students would not add the unlike denominators together because the incorrect solution of twelfths (4/12) could not be reasonably explained through a drawing.

Comparing and discussing

a) Begin with a student who got stuck (R3) to clarify what makes the problem difficult.

b) R2: ¼ in half to create two ⅛ pieces and so can add ¼ + ⅜ and find that ¼ + ⅜ is ⅝

c) R4 shares how they used a number line with both ¼ and ⅛ intervals on it to find equivalent fractions to make sense of how to add fractions with unlike denominators.

d) Conclude with R1:

$$1/4 = (1 \times 2)/(4 \times 2) = 2/8$$

so

$$¼ + 3/8 = 2/8 + 3/8 = 5/8$$

In facilitating the discussion, both the teacher and the other students will ask questions of the student that shares R2 to get them to recognize that when they decompose ¼ in two pieces, those two new pieces are each ⅛, so their number sentence should reflect that. They did not actually add ¼ + ⅜; they changed ¼ into the equivalent fraction 2/8 so the number sentence should be 2/8 + ⅜ = ⅝.

Summing up

Teacher records the summary, as dictated to her by students: "Today as a hard-working class, we learned that we can add fractions with unlike denominators by using equivalent fractions."

Board plan

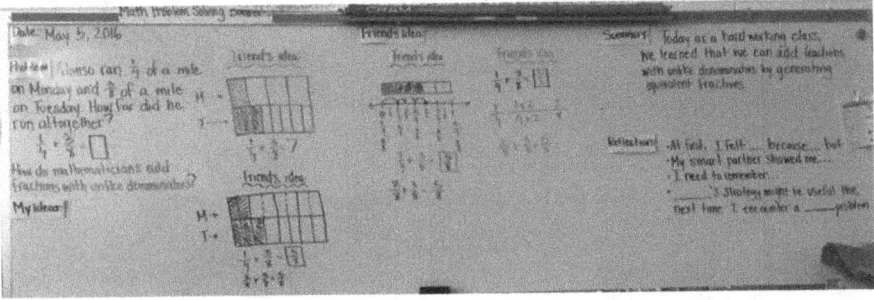

Figure 5.8 The board during the planning stage demonstrates the anticipated student responses

Observations from the lesson

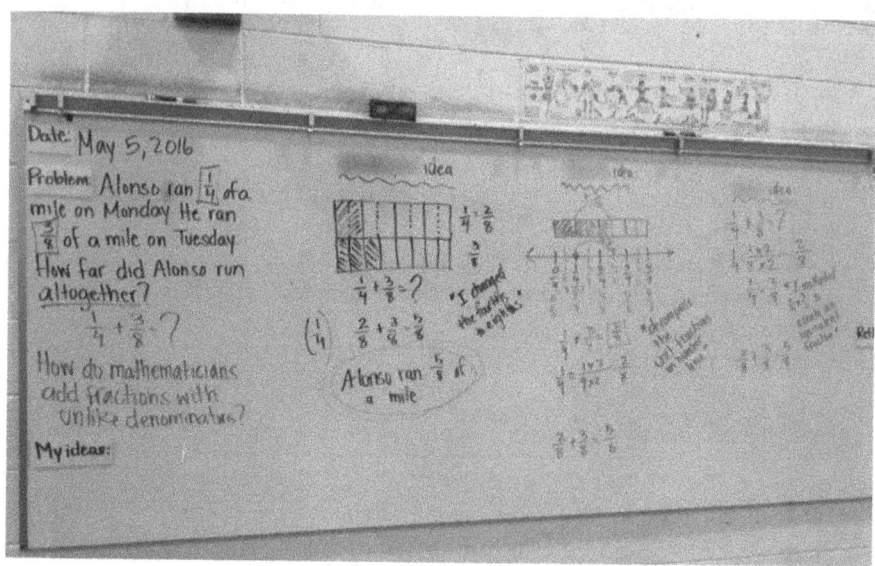

Figure 5.9 The left side of the board after the lesson illustrates three students' strategies, ranging from pictorial to abstract

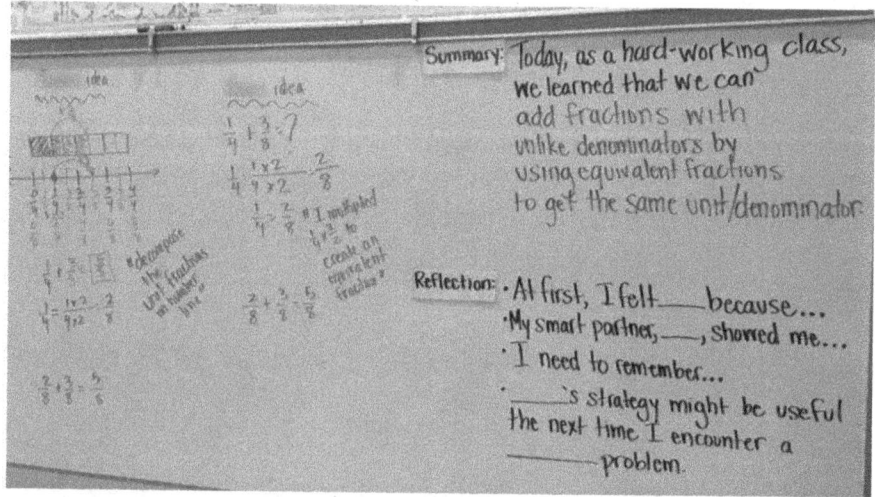

Figure 5.10 The right side of the board depicts the student-created summary at the end of the lesson

Some common themes that observers noted included how the classroom culture, lesson structure, routines, and prior mathematical experiences clearly contributed to the success of this lesson. In addition, many observers noted the influence of Prieto's years-long emphasis on student discussion. Further, Ms. Johansen's class was an example of the school-wide research theme, to help students *construct viable arguments and critique the reasoning of others* – CCSS-Math Practice 3 (Common Core State Standards Initiative, 2010). Post-lesson small groups noted that in this lesson, the students did the heavy cognitive lifting, while Ms. Johansen was there to facilitate the deep discussion of their ideas. This was made possible through the strategic balance of instructional time spent in various modes of independent and collaborative work.

Observers also noted that in this lesson, students were able to quickly draw out their own *Guiding Question* (i.e. objective), "How do mathematicians add fractions with unlike denominators?" from the story problem. Traditionally, the teacher simply tells students the objective at the beginning of the lesson. In this lesson, however, students were able to ascertain the objective themselves by analyzing the story problem and then applying it to a broad mathematical concept.

Final comments and keynote speech

Dr. Akihiko Takahashi remarked that he was impressed with how much the fourth grade students in this class carried the weight of the lesson, despite it technically addressing a fifth grade standard. He noted how well students listened and responded to one another's ideas in turn and talks, table conversations, and during whole-group discussion. He also acknowledged that many students approached the problem through various strategies, continuing to find creative methods for solving, even after they had arrived at their first solution. That is, many students had clearly understood the value of proving one of their ideas (e.g. an algorithmic approach) with another idea (e.g. a fraction model approach). In turn, Dr. Takahashi mentioned the clear connection between the students' notebooks and the teacher's board writing. The whiteboard illustrated student ideas from pictorial to abstract, allowing for multiple access points to the problem. Further, Dr. Takahashi explained the important distinction between "answer-getting" and permeating mathematical learning. Thus, the *Summary* of the lesson was not the answer to the story problem (*How far Alonso ran*), but rather the answer to the Guiding Question (*How to add fractions with unlike denominators*).

Because Ms. Johansen's public research lesson was taught at a large conference (LSRG, 2021), there was also a keynote speaker, Dr. Alan Schoenfeld of UC Berkeley, who remarked on the lesson after Dr. Takahashi. Dr. Schoenfeld framed his comments within the structure of his Teaching for Robust Understanding (TRU) Framework (Schoenfeld, n.d.). He discussed how the mathematics was able to flourish because of the evidently internalized discussion routines of the class. He also spoke about how the structure of the lesson gave students *Agency, Authority, and Identity* (i.e. ownership) over their learning experience.

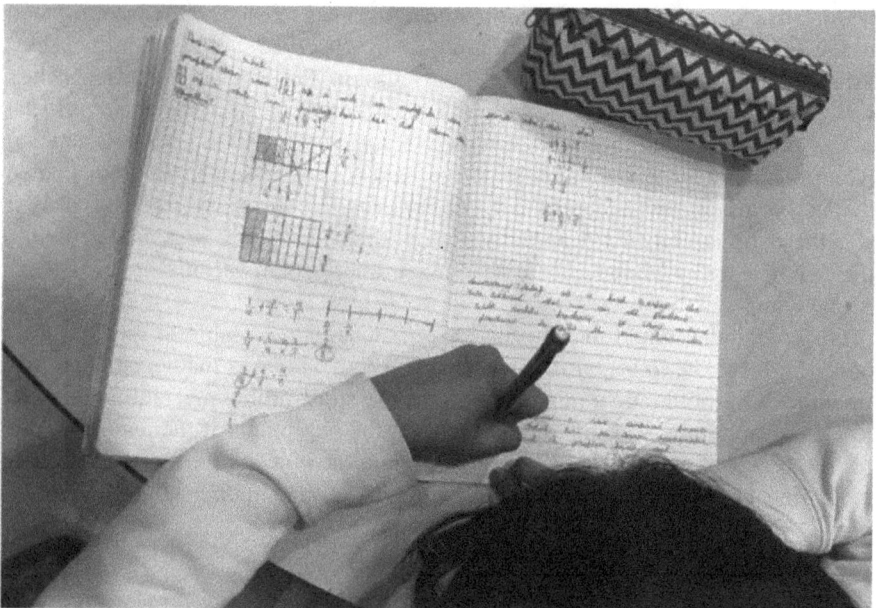

Figure 5.11 A student grapples with various approaches, giving her several different
solutions

Impacts on my own work and perspective

As a first-year teacher in 2014, I was immersed into the world of education with
little prior experience teaching elementary mathematics. While I had previously
gained some classroom experience teaching high school English as a Foreign
Language (TEFL) in Madrid, Spain, nothing could have prepared me for the challenges of teaching all subjects to elementary students in the very different context
of Chicago Public Schools (CPS). I had the good fortune to teach fourth grade
at Dr. Jorge Prieto Math and Science Academy, with its tightly knit community
of families and students who were excited to learn. I was also able to lean on a
supportive team of colleagues and an administrative team that understood the
value of student voice. Nevertheless, I still had a lot to learn about high-quality,
transformational instruction. Throughout my tenure in CPS, the bulk of my pedagogical understanding of strong mathematics instruction came from the Lesson
Study process and through practicing Teaching through Problem Solving (TTP).

During the summer leading up to my first year of teaching in CPS, there was no
shortage of advice offered to me by veteran teachers, content coaches, and school
administrators. Professional development opportunities were flying in my direction
from all angles. I attended every opportunity that was presented to me and began the
process of sifting through the heaps of presentations, seminars, and PowerPoints to
try to figure out my role in creating equitable access to opportunity for my students.

My first experience learning about "constructivist" pedagogy took place during a mathematics methods course I took just before I began teaching at Prieto. The course was centered around a lesson structure called "Launch, Explore, Discuss" whereby students were presented with a problem in the context of a story, given time to investigate various potential approaches to solving the problem, and then asked to come back together as a class to discuss classmates' various ideas. While the process made sense to me in theory, I had no idea what this would look like within the context of my future classroom. What I know now is that effective professional learning involves continuous study, ongoing observational feedback, and structured collaboration with colleagues. Enter: Lesson Study.

Both my future students and I got lucky in the summer before my first year, when I was presented with the opportunity to attend a four-day professional development workshop called Lesson Study. At this workshop, I heard Professor Akihiko Takahashi talk about the value of an approach called Teaching through Problem Solving (TTP), which went beyond what I understood about the "Launch, Explore, Discuss" model in that there was a defined learning goal that was to be articulated (and hopefully demonstrated) by the end of the lesson. Also at this four-day Lesson Study workshop, we watched videos of teachers demonstrating their understanding of the value of giving students the structured space to independently grapple with new math concepts.

A few months after the workshop, my grade-level colleagues and I were invited to attend the teaching of a research lesson, during which one of the members of a planning team taught the team's lesson in front of observers. I remember thinking how absolutely terrifying this would be and noting that I would never teach a public lesson like this. Nevertheless, during my second year at Prieto, my grade-level team and I began planning a fractions research lesson, and after some discussion among our team and with school administrators, it was decided that I would teach the public research lesson in May of that year.

As a second-year teacher, the idea of teaching a lesson I had never taught before in front of 200 educators, school administrators, and district leaders was unsettling. However, through the process of Lesson Study, I was given a framework for collaboration and a built-in support system of co-planners with whom I could hash out every detail of the lesson. Moreover, my colleagues and administrators believed in me, and I prepared more for that lesson than I had even for college exams, and took pride in doing so. The lengthy discussions about the intricacies of fractions made my brain buzz like it hadn't since I myself was in school, and I began building on the foundation of seeing math as a tool for creating a classroom focused on both deep understanding of mathematics and equity. I had always heard that teaching was not the "filling of a bucket, but rather the lighting of a fire", and now I had a support system to help me understand the pedagogical power of Teaching Math through Problem Solving (TTP) in lighting that fire. Creating transformational change for my students and giving them equitable access to learning opportunities has always been my "why" and my "what". Now, I had TTP, which for me and my students was the "how". That is, *how do we help students find joy in experiencing the beauty of mathematics and its connection to the real world? How do we help students to really listen and build off of one another? How do we give students agency over their learning experience? How do we empower students with habits of mind that set them up for future success in life?*

Admittedly, in the midst of academic-style conversations about the nitty-gritty details of student understanding of elementary mathematics, I've wondered to myself, "in the grand scheme of equity and creating opportunity, where does all of this fit in?"

My own theory and answer to the previous question involves several components:

– when one approaches teaching pedagogy through the lens of equity, every small component of classroom culture, teacher response, student thinking, and student voice matters. That is, we should "sweat the details".

– not unlike students, professionals also rise to expectations when given strategic supports, encouragement, and feel that the work is important.

– while there, of course, is value in students understanding the mathematical principle that two unlike units can be converted to the same unit to be combined; for me, the real, permeating lesson centers around students' own belief in themselves to self-actualize and direct their own paths forward into a future of choice and opportunity.

Note

1 A full version of the lesson plan can be accessed at https://LSAlliance.org/Lessons.

References

Common Core State Standards Initiative. (2010). *Common Core State Standards for Mathematics*. Retrieved from www.corestandards.org/Math/

Schoenfeld, A. (n.d.) *Teaching for Robust Understanding Framework*. Retrieved July 24, 2021, from https://truframework.org/

LSRG (2021). *TTP in Action*. Retrieved July 24, 2021, from https://lessonresearch.net/teaching-problem-solving/ttp-in-action/

Building district infrastructure for Lesson Study: designing educator support centering our students at the margins

The San Francisco Unified School District (SFUSD) Teacher Leader Fellowship (TLF) and Lesson Study support programming

Nora Houseman

In 2008, the voters of San Francisco passed a proposition (Prop A: Quality Teacher and Education Act), funded by a parcel tax. Amongst other goals, this proposition called for the development of an SFUSD Master Teacher Program "in the interest of providing a pathway for teacher leadership as well as support for selected school sites" (*UESF, 2017, p. 157*). The school district and the educator union (United Educators of San Francisco) then began collaborating to "develop a Master Teacher program and provide incentives for exceptional teachers to stay in the classroom to support student achievement and promote professional learning communities" (*UESF, 2017, addendum*). The current program supervisor at the time, Stephanie Ervin, seeking innovative means to test and refine new Common Core curricula initiatives in SFUSD, learned of the Lesson Study work at Mills College (Drs. Catherine Lewis, Shelley Friedkin) to support the teaching of mathematics through Lesson Study. Stephanie visited Mills, became inspired by the possibilities of Lesson Study, co-designed the Master Teacher Program with the support of Catherine and Shelley, recruited 20 teachers willing to try Lesson Study, and launched the early iteration of the Master Teacher Program.

Since this time, the program has grown to 45 teachers, evolved its purpose, and adapted financially and programmatically to changing times to remain transformational and relevant. At the heart of the program, however, remains a commitment to Lesson Study.

This section will explore how a public school district utilizes Lesson Study as a means of improving the experience and learning outcomes for all students, and particularly historically underserved students, through:

- an overview of the Teacher Leader Fellowship (TLF) – formerly called the Master Teacher Program

DOI: 10.4324/9781003230915-28

- an analysis of how the Teacher Leader Fellowship supports transformation
- a reflection on how the program has adapted to shifting priorities and addressed challenges to maximize impact
- sharing of current thinking on the potential of our model.

The SFUSD Teacher Leader Fellowship

The SFUSD Teacher Leader Fellowship is a four-year teacher leader development program. Any SFUSD teacher can apply (any role, any grade level) with a minimum of four years of teaching experience. Applicants must demonstrate – through a rigorous application, interview, and observation process – strong classroom practice, a commitment to further equity through teacher leadership, and a willingness to recruit and lead a Lesson Study team at their site, focused on instructional improvement and teacher ownership. Once accepted, Teacher Leader Fellows (TLFs) commit to:

- attend monthly three-hour meetings with all TLFs (cross-school professional development)
- lead two cycles of Lesson Study with research lessons
- support site-based professional learning and teacher leadership
- formally present learnings at end of year.

TLFs are supported and compensated to accomplish the previous with a $2500 stipend; 25 hours of pay and two sub days for the TLF and each member of their Lesson Study team; hourly pay for attending the TLF monthly meetings; and an assigned coach that provides monthly leadership coaching and Lesson Study support through a gradual release model. Coaching includes tools for each step of the Lesson Study process, support crafting a compelling research question and engaging in the research cycle, support designing and implementing Lesson Study according to the needs of the site, facilitation at research lessons, guidance for creating and tracking a leadership goal, and support problem-solving leadership/facilitation challenges that arise.

All TLFs meet monthly for three hours at the cross-site Professional Learning Community (PLC). The PLC has three purposes for TLFs:

1   *Develop/hone an equity lens specific to public education in the SFUSD context and commit to interrupt, transition, and transform inequities in context – personal and professional.*
2   *Experience, facilitate, and learn best practices for PD/PLCs (rites and rituals, structures and protocols, responsive facilitation) in order to:*

   A   *Grow as facilitators of PD and of Lesson Study (by facilitating and receiving feedback)*
   B   *Deepen our leadership content knowledge; develop a toolbox (i.e. protocols and structures, responsive facilitation moves) to hone our skills and*

strategies to "hold space" for individualized, equity-centered learning in our Lesson Study teams, classrooms and schools

    C   *Experience and develop an awareness to who and how we – in the skin we are in – exist in our work in relation to the reproduction of oppression and privilege in education*

3   *Provide a rich learning space for teachers to build alliances and learn from their colleagues across site, role, and difference.*

At the PLC, TLFs engage in community building, text protocols (usually with a selected book for the year), racial affinity groups, dilemma protocols, leadership development, and the sharing and tuning of Lesson Study practice. TLFs also have opportunities to practice and receive feedback on equity framing, process observing, and facilitating protocols.

Because of the Lesson Study network, the program is able to support and facilitate meaningful collaboration across sites and Lesson Study teams. This includes connecting schools with similar research themes to share learnings, supporting TLFs to attend research lessons or observe Lesson Study meetings at other sites, organizing and facilitating cross-site research lessons (teams with participants from multiple schools) that teach publicly for our district, sharing unit and planning templates across sites, and connecting TLFs with similar leadership goals or dilemmas to learn together.

The TLF is run by a specified team within the division of Curriculum & Instruction (called the Professional Learning & Leadership Team). This team has a supervisor and roughly four teachers on special assignment that design the program, facilitate the PLC, coach TLFs, and support Lesson Study teams.

How the TLF supports transformation

Program Vision:

The SFUSD Teacher Leader Fellowship strives to identify and develop strong teachers committed to equity-centered leadership, in order to:

■   *Support teachers to create, foster and sustain equity-centered professional learning communities that will transform individual teacher practice and increase student success – impacting both student achievement and student experience.*

■   *Support teachers to develop as responsive facilitators for equity and social justice, able to lead professional development and change initiatives at their sites that are contextual, responsive, and strategic.*

Toward the goal of: Ensuring all students have access to personalized, equitable and high performing schools that believe and demonstrate each student can, should and will succeed.

And ultimately: Interrupting and transforming current and systemic educational inequities through a belief in grassroots change (ground up).

As evident from the program vision, the TLF and Lesson Study are intended to be levers for individual, site, and district transformation. Participating teachers and school sites have experienced significant success – in student achievement data; in ability to grow teacher practice, expertise, and leadership; and in shifting the learning culture of schools. This is evidenced through TLF and Lesson Study participant feedback and self-assessments of growth; qualitative and quantitative data collected and measured through Lesson Study cycles; and site and district achievement data demonstrating growth over time and by comparative cohorts.

Unlike other Lesson Study models, the SFUSD model explicitly prioritizes our students most in need, with the goal of closing the opportunity gap for historically underserved students. TLFs are supported to focus on the learning of their identified focal populations (this may be Black, Latinx, American Indian, or Pacific Islander students; English language learners; students with special needs; or any other population identified as "least served"). In the PLC, this may take the form of internal identity work or racial affinity group work with application to shifting practice to better support focal students; focal student tracking and analysis protocols; or learning and applying leadership and teaching practices to better meet the needs of focal populations. During site-based Lesson Study, this may look like engaging in listening and co-designing protocols with focal groups at all stages of the process; conducting research specific to the focal population; designing lessons explicitly to meet the needs of focal students; collecting implementation and impact data of focal students; or scripting and observing focal students during research lessons.

We believe that instructional practices supporting those most in need will inherently support all students. We are therefore as concerned with focal student data as we are with overall class or cohort data. We have observed Lesson Study models in which success is attributed to generalized data collected ("80% of students were able to . . .") that, if disaggregated, would reveal concerning disparities in how students experienced the lesson by gender, race, language, and economic background. Our research model therefore zeros in on the experiences and learning of focal students by ensuring that the data/observations collected, the post-lesson discussion, and any consideration of next steps all center the focal populations identified.

Additionally, our Lesson Study model is not intended to be performative, in that the goal is not to "deliver" a flawless research lesson. Rather, we hope that the Lesson Study process and the research lessons are genuine opportunities to experiment with research-based and field-tested practices that the team has identified as "promising" for their focal populations, engage in analysis, and receive critical feedback. We encourage "messy" research lessons, and steer educators away from scripting and instead toward innovating on the spot – considering multiple pathways a lesson may take, and coming prepared to follow any of these paths – or a new direction altogether if that is what naturally evolves. We believe this supports the learning of focal students because this model encourages designing lessons *toward* the needs of focal students, rather than deprioritizing the needs of focal students to design for the most advanced students (as public performance), or, in the worst-case scenarios we have observed, removing focal students from the room so that a research lesson might proceed more "smoothly".

How the TLF has adapted

To maintain district-wide support, the Teacher Leader Fellowship has remained continuously agile, adapting as needed to meet shifting priorities and address arising challenges. A few examples of such adaptation will be shared.

Example A: moving from a team model to a schoolwide model

Five years ago, we identified that while Lesson Study was impacting the practice of individual team members, most Lesson Study teams operated in a bubble, with minimal school interaction and impact. We therefore created a "whole school Lesson Study process", working with existing TLFs and their administrators to recruit groups of TLFs from the same site, and providing more flexible support to these schools as we co-designed a school-wide Lesson Study model customized to the needs of each participating site. This process was enabled with the support of our Mills colleagues, who through a continued grant, connected us with Lesson Study teams and observation opportunities in Oakland, Chicago, and Japan; provided consultation and direct team support; and worked with us to grow our own practice to support teams, schools, and networks. Five schools initially prioritized Lesson Study as their primary PD model, and we worked collectively to design structures for teams within each school to share learnings as well as structures for how the five schools could learn from one another. We experimented with many formats for how teams within a school might operate as a collective: beginning school-wide with a research question and whole school research, then splitting into grade-level teams to design lessons and focus on developmentally specific practices; operating as individual teams but meeting whole-school periodically to engage in artifact shares and feedback/tuning protocols; operating as individual teams but coming together whole-school for research lessons, with time built in for grade-level bands to consider implications and next steps; creating leadership teams that wrote newsletters to share the work of each team schoolwide and/or met regularly to design and align their Lesson Study meetings and goals. We also began meeting quarterly as a network of whole-school Lesson Study sites, to learn from the successes and challenges of each site and share best practices.

While this work had considerable success, the financial and time challenges of whole school Lesson Study remain a barrier to growth. For the model to succeed in our district, schools need to devote ample time to planning, be willing to set aside other district priorities to focus on Lesson Study, continuously grow or maintain strong teacher leaders, and creatively find ways to hold research lessons when substitutes and release time are rarely reliable or financially feasible. We continue to work with those schools invested, but given rates of teacher and administrator turnover in our district, few schools are ready to make the long-term commitment to whole school Lesson Study. In the coming years, we hope to create clear district-supported pathways for Lesson Study sites, so that the barriers of entry minimize and sites can select into networks that enable, rather than complicate, the implementation of Lesson Study.

Example B: ensuring district-wide leadership and impact

The TLF proactively seeks to be at the forefront of district initiatives. In recent years, our district created a new Graduate Profile and articulated a move toward Deeper Learning and antiracist practice. To align with these goals, we formalized our equity work within the PLC, launching racial affinity groups and selecting texts with explicit themes of antiracist leadership. We re-wrote our Lesson Study tools, ensuring equity at the center and a focus on focal populations. We also began recruiting TLFs invested in project-based learning (PBL), and eventually built a cohort large enough to launch a TLF track focused primarily on PBL. We now operate two cohorts in the program, each with 12–25 TLFs, with the PBL cohort modeling paving the way for what a transformative SFUSD model of PBL looks like in practice in our schools.

Finally, the TLF seeks to have district-wide impact, despite serving a small number of teachers within a large district. When faced with critique that our program was "boutique" and unscalable, we began reimagining how we might maximize impact. We began hosting "district-wide public research lessons", open to all educators, administrators, and district leaders – as well as external visitors. This was a means to share the work and learning of exceptional Lesson Study teams, but also a means to hold district-wide conversations about promising practices and demonstrate new ways of envisioning professional learning and teacher leadership at the site level. We also began hosting professional learning for broader networks of schools, inviting educators to engage in PD cycles about promising practices evolving from our Lesson Study schools, such as Teaching through Problem-Solving in mathematics. We continued to prioritize teachers working with our least served populations, targeting our program recruitment and application process toward these teachers. As we built our whole school Lesson Study network, we selectively partnered with schools on our southeast side (schools serving historically underserved youth), ensuring our investment and resources were directed to those schools with most need. Finally, we began more intentionally collecting data on our program outcomes (teacher retention and agency; student experience and achievement; school culture change) and publicizing our successes widely – to district leadership, school board members, central office staff, and school staff. All of this has supported the growth in reputation of our program, positioned us at the forefront of pedagogical practice, and allowed our work to permeate beyond program participants. That said, our goal to have district impact is ever shifting, and we continue to consider what moves to make each year, as we assess the landscape of a district faced with annual budget cuts, ongoing changes in leadership, and continued disparities in student achievement.

Potential of the model

We believe the Teacher Leader Fellowship is a unique and exceptional program, with the potential to serve as a blueprint to shift educator practice and teacher leadership in school districts nationwide.

Unlike other school change models, the TLF is a grassroots program, focused on educators working directly with students, that does not rely on administrator or district leadership to succeed. This is rare in a nation in which school districts continue to be organized hierarchically, with most pushing top-down initiatives that rarely "catch" at the classroom level. We believe there are revolutionary implications for centering educators in school change, positioning teachers as the experts of their own inquiry cycles, and supporting teacher agency through continuous improvement processes that occur at the classroom level and demand change *upwards*.

Further, our model rejects compliance and standardization, and centers creative adaptation and the humanity of the educators and students involved. Each TLF designs their own Lesson Study process, customized to the needs and specifics of their site and their students. There is no "one-size-fits-all" model, and our coaches are skilled in thinking outside the box and supporting thoughtful experimentation in each unique context. We are careful not to encourage blind replication of structures, tools, or processes – but instead to rely on the expertise of teachers to innovate in service of their focal populations. We are tight in our values and in ensuring use of best practices within inquiry cycles, and loose in the specifics of how to cater Lesson Study to each school context.

Despite all the challenges, we remain deeply committed to Lesson Study as the central focus of the Teacher Leader Fellowship. For us, Lesson Study is not just a PD structure, but an intentional choice that interrupts the focus on discipline and control so prevalent in schools serving Black and Brown youth, and focuses educator energy on the learning of our least served students. It is an intentional choice toward academic justice and liberation, in a nationwide education system in which opportunity gaps persist and inequity by design prevails. We choose Lesson Study because it attends to the intellective capacity of each and every student, and places the responsibility for learning on the system – rather than demand that students conform to an inequitable system that has historically denied access, opportunity, and humanity to our least served youth. In short, Lesson Study continues to serve as the anchor of our program model because it enables, in process and product, progress toward the liberating learning spaces we envision for each and every student in SFUSD.

Reference

UESF. (2017). *Certificated collective bargaining agreement, July 1, 2017 – June 30, 2020*. https://uesf.org/wp-content/uploads/2018/04/Certificated-Collective-Bargaining-Agreement-7–1–17-thru-6–30–20-pre-final.pdf

Summary of Chapter V

What we learned about teacher collaboration and leadership

Shelley Friedkin

In this chapter, we learn how collaboration throughout Lesson Study provides the opportunity for team members to express, explore, challenge, and evaluate their thinking and professional practice. We also understand how teams build professional trust and growth mindsets as they move through the Lesson Study cycle, and as they publicly share their hypotheses and practice about teaching and learning through research lessons. A lens into how shifts in teaching and learning occur at both an individual and team level is revealed, and we come to understand how this builds the path for school-wide instructional improvement. We also learn about a cohesive district model of Lesson Study where the complexities of teacher collaboration and Lesson Study facilitation are examined and leveraged to create meaningful teacher leadership opportunities. This district model also exemplifies how Lesson Study can be used to deepen a district's equity vision. Collegial learning, shifts in pedagogical practices, a deeper understanding of content progression, the gradual and authentic development of leadership, self-reflection, and professional confidence in decision-making become apparent as we hear from our authors about their meaningful Lesson Study collaborations.

As Cassie Kornblau begins her Lesson Study report focused on decimals, the energy she describes in the room where the team prepares for their research lesson is palpable. The collective focus, the level of detail, the collegial support, and the continued strive for excellence to benefit students' sense-making can be heard, seen, and felt. Kornblau described how close the team had become as colleagues, and how willing they were to argue, discuss, and push each other's thinking. She also reflected on the cultural shift at her school, where teachers pushed forward their own learning as they led each other through the process of Lesson Study. Lesson Study, Kornblau articulates is 'a place where teachers can slow down, look closely at what they are teaching, how they are teaching it, and most importantly, why.' The collective learning from Kornblau's team inquiry revealed the need for school-wide improvement around the conceptual progression of decimals and numbers in general. More specifically, the team highlighted that students needed more time across the grade levels with one-dimension measurement and number placement.

Brigid Brown's report focuses on quotative division and, similar to Kornblau, shows how across grade-level Lesson Study team members gained an

DOI: 10.4324/9781003230915-29

appreciation of other grade-level perspectives. Through their Lesson Study work, Brown's team solved their school's long-held problem of practice associated with students' lack of understanding when solving word problems. The team's inquiry focused on how to work with students to unpack the language of the word problems effectively and to make use of models and equations to calculate, all without guiding students through each step of the process. As the team collectively examined student learning from a range of grade-level classrooms and worked through their Lesson Study cycle, it became apparent that a robust understanding of word problems goes hand in hand with students' ability to draw and label a diagram to match the word story and express connections between the quantities and relationships in the story problem. Brown shared as a result of this collaborative work 'teachers embraced math instruction more deeply, many saying that they enjoyed teaching math for the first time in their careers. And our students began participating eagerly in challenging classwork and naming math as their favorite subject.'

As a new teacher, Alex Johansen Laughlin shares the complexities and challenges of teaching students how to add fractions with unlike denominators. Through the process of Lesson Study, she experienced a strong framework for collaboration and a built-in support system of co-planners to hash out every detail of the research lesson and student thinking. She recalled how her learning fell into three main buckets: mathematical, pedagogical, and professional. Mathematically, she learned that teachers need to create opportunities for students to see connections between mathematics and life. Also, fractions are challenging to teach because of their many interpretations and how students experience confusing introductions to fractions. Pedagogically, she learned that students thrive in a predictable lesson structure and how students' discourse norms support various ways of student thinking. Professionally, she learned that teachers (just like students) rise to expectations when given strategic support and encouragement and when they feel the work itself matters. Alex Johansen Laughlin, as a second-year teacher, and with the collegial support of her Lesson Study team, taught a public research lesson observed by over 200 audience members.

Nora Houseman provides a detailed lens into how San Francisco Unified School District's (SFUSD's) Professional Learning and Leadership team built an infrastructure for Lesson Study to grow teacher leadership, and more recently to formalize the district's equity vision and anti-racism. SFUSD's Lesson Study program has protected funding to provide comprehensive and ongoing support for Teacher Leader Fellows (TLFs) who facilitate the collaborative learning spaces for site-based Lesson Study work. Houseman shares how support for TLFs include monthly meetings where they learn about community building, facilitation, dilemma protocols, leadership development, to name a few, along with ongoing opportunities to share, tune, and receive feedback around their Lesson Study practice. TLFs dive deeply into who they are as educators and how they show up in relation to the reproduction of oppression and privilege that exists in education.

Authors in this chapter collectively surface how Lesson Study collaboration involves purposeful actions to enhance their understanding of teaching and learning and impact outcomes for students. They also expose how deep relationships between team members develop autonomy, belonging, and competence as they expand their knowledge and strengthen their beliefs and habits about teaching and learning. Participation in their Lesson Study communities allowed each educator to build their identity as a teacher, a learner, and a leader simultaneously.

VI Ideas for establishing sustainable Lesson Study – what we learned from US schools

DOI: 10.4324/9781003230915-30

What we learned from twenty years of U.S. schools' endeavors

Akihiko Takahashi

The beginning of Lesson Study outside Japan

What we know as Lesson Study comes from the Japanese practice of *Jyugyou Kenkyuu*. *Jyugyou Kenkyuu* became a popular professional development approach outside Japan after U.S. researchers shared the success Japanese teachers had been having with it at their schools. Case studies (Mills College Lesson Study Group, 2000, Lewis, 2000; Lewis & Tsuchida, 1998; Stigler & Hiebert, 1999; Yoshida, 1999) highlighted what Japanese teachers, particularly mathematics teachers, do in and outside of the classroom and how these teachers improve their teaching and learning. Despite international assessments, such as the Second International Mathematics and Science Study (Chang & Rusicka, 1985) and the Third International Mathematics and Science Study (Mullis et al., 1997), which showed that Japanese students performed better in mathematics compared with their peers worldwide, only a few studies had been conducted on how Japanese mathematics teachers teach and how their students learn. Some of these early studies about Japanese mathematics teaching and learning include *Some Observations of Mathematics Teaching in Japanese Elementary and Junior High Schools* (Becker, Silver, Kantowski, Travers, & Wilson, 1990) and *The Learning Gap: Why Our Schools Are Failing and What We Can Learn from Japanese and Chinese Education* (Stevenson & Stigler, 1992). These reports may have motivated U.S. researchers to conduct more in-depth studies.

Building onto these studies, a team of U.S. researchers conducted the TIMSS Videotape Classroom Study (Stigler, Gonzales, Kawanaka, Knoll, & Serrano, 1999), which used classroom videos of eighth-grade mathematics from the U.S., Germany, and Japan to compare and contrast their teaching approaches. This led to the publication of *The Teaching Gap: Best Ideas from the World's Teachers for Improving Education in the Classroom* (Stigler & Hiebert, 1999), which was based on the TIMSS Videotape Classroom Study. Researchers and educators worldwide, particularly in the U.S., became interested in learning about Japanese mathematics teaching and learning. *The Teaching Gap* also describes how the Japanese school system uses *Jyugyou Kenkyuu* as the primary form of professional development for practicing teachers.

DOI: 10.4324/9781003230915-31

Immediately after the publication of *The Teaching Gap*, the International Congress on Mathematical Education (ICME-9) was held in Tokyo in 2000. Two seminars were then organized by the U.S. and Japanese delegates to further discuss Japanese mathematics instruction. One was organized by the Mathematical Sciences Education Board and U.S. National Commission on Mathematics Instruction (National Research Council, 2002). The other was organized by the National Council of Teachers of Mathematics (Curcio, 2002). Key U.S. and Japanese mathematics education researchers participated in either one or both of these seminars. The purposes of these two events were almost identical: to learn more about how Japanese teachers teach mathematics and how they develop the knowledge and expertise for teaching math through the practice of *Jyugyou Kenkyuu*.

One of the significant discussions in the seminars was to clarify the definition and operating principles of *Jyugyou Kenkyuu*. Although the Japanese have used Lesson Study for more than a hundred years, even the Japanese mathematics education researchers and teachers did not have a consensus on how to define *Jyugyou Kenkyuu*. For example, Hirabayashi, a leading mathematics education researcher who participated in one of the seminars, described *Jyugyou Kenkyuu* as a "method of research in mathematics education" (2002). However, most U.S. researchers thought of *Jyugyou Kenkyuu* as a form of professional development for teachers, a definition with which some Japanese researchers also agreed. Two English translations of *Jyugyou Kenkyuu* were also offered at the time. One, "Lesson Study," was proposed by Makoto Yoshida (1999), and the other, "Lesson Research," was provided by Catherine Lewis (2000). Since the concept of "teacher research" was not common outside Japan at the time, researchers and educators generally adopted the term "Lesson Study" as a way to refer to *Jyugyou Kenkyuu* as the traditional form of professional development in Japan.

Limitations of early attempts of using Lesson Study in U.S. classrooms

Soon after Lesson Study was introduced outside Japan, many researchers tried to replicate it in U.S. schools. Since most U.S. primary and secondary schools did not have resources for professional development, researchers had to secure external funding. Thus, most of the early Lesson Study projects in the U.S. were planned and led by outside researchers and professional development specialists who brought external funding to schools and districts. They secured extra funding from federal or private foundations and recruited volunteer teachers to try Lesson Study as an additional activity to their daily teaching. These researchers reported some advantages of using Lesson Study in this way to support teachers and students in improving teaching and learning. Still, the outcomes of such ad-hoc professional development programs were often limited (e.g., Fujii, 2014; Hart, Alston, & Murata, 2011; Takahashi, 2011). Although most Lesson Study projects were designed to establish sustainable teacher support structures for teacher-led professional development, the same as in most Japanese elementary schools, teachers and educators often decided not to continue after the conclusion of their external funding and support (e.g., Akiba & Howard, 2021; Lewis, 2002a; Takahashi & McDougal, 2016).

Many other countries also tried to use Lesson Study to update teaching and learning. In addition to the U.S. researchers and educators who introduced Lesson Study to countries outside Japan, Japanese researchers and educators also began to spread the practice of Lesson Study to other countries. These Japanese researchers and educators often brought experienced Japanese teachers to classrooms in other countries to demonstrate Japanese ways of teaching with the local students using simultaneous translation. Such demonstration teaching may help local teachers see how their own students engage in learning led by Japanese experts. However, it often brings some confusion about what Lesson Study is. The demonstration lessons may show what the Japanese teachers accomplished through Lesson Study. However, the entire process and ideas behind the Lesson Study are often not visible from just watching a demonstration lesson. These early attempts may have hinted that the Japanese Lesson Study may not be easily adaptable as is outside Japan because Japan's school culture and society are so different from that in other countries.

Making Lesson Study school-wide

Among the U.S. schools where Lesson Study has been piloted, some schools have shown substantial impact on teacher and student learning (Lewis, 2016; Lewis & Perry, 2014; Lewis et al., 2006; Mills College Lesson Study Group, n.d.; Takahashi et al., 2013; Takahashi & McDougal, 2016). One of the keys to establishing sustainable Lesson Study is to have a teacher leader(s) in the school to encourage other teachers to participate in Lesson Study. These teacher leaders often lead their colleagues to experience Lesson Study as an opportunity to learn from each other. Soon other teachers see the benefits of Lesson Study, and it becomes a school-wide endeavor for the teachers at the school to address issues in improving teaching and learning.

The beginning of these schools which established school-wide Lesson Study was similar to the other schools where Lesson Study ended when the funding did. Enthusiastic teachers worked with outside experts of Lesson Study to experience a full Lesson Study cycle: they developed lesson proposals, conducted research lessons, and reflected on the process to summarize what they had learned. Through this process, these teachers learned that Lesson Study is a way for them to develop shared ideas for improving teaching and learning at their school. The teachers realized they can work together to produce a shared understanding of their students' needs, the curriculum, and pedagogical ideas.

Once the teachers recognized that Lesson Study helps them investigate theory and research how to improve teaching and learning in their classrooms, they were excited to share their knowledge with their colleagues. The core Lesson Study teachers gradually began to invite other teachers to join their research lessons. After a year or two of Lesson Study, other teachers joined the lesson designing team. The core teachers who participated in the initial Lesson Study cycles gradually became leaders of Lesson Study at their schools.

This gradual inclusion of many teachers seems critical among the schools where sustainable Lesson Study is achieved (Takahashi & McDougal, 2016). As more teachers get involved, the teacher leaders communicate with their school administrators to allocate more time and resources to Lesson Study in order to establish it as the school's major professional development activity. This allows other teachers interested in joining to be a part of the endeavor. Eventually, it becomes a large enough group that the teachers can set up a research steering committee and choose a common research theme for each Lesson Study cycle (Takahashi & McDougal, 2016).

The process for establishing a school-wide Lesson Study often takes more than three years. After it becomes fully school-wide, the schools start showing visible change among teachers and students. For example, John Muir Elementary, a school in the San Francisco Unified District, has shown significant improvement on student assessments as compared to other schools in the district not using Lesson Study (Figure 6.1). Although more study is needed to unpack how this improvement happens, the teachers at these schools can already see the results of their tireless efforts.

Beyond professional development

Although Lesson Study was introduced as a form of professional development, some pioneers of Lesson Study outside Japan realized that it has unique characteristics that set it apart from a typical professional development program. Lewis (2002b) contrasts Lesson Study with traditional professional development programs, based on an idea by Lynn Liptak, one of the first principals who endorsed Lesson Study (Table 6.1). As Table 6.1 shows, Lesson Study allows teachers to begin with a question and then investigate that question in their classrooms, the place where the elements of the instructional triangle (National Research Council, 2001) – student, teacher, and mathematical content – emerge. Thus, through Lesson Study, teachers can produce specific knowledge and expertise that allows their students to learn better. This may be why Hirabayashi (2002) emphasized Lesson Study as a method of research for teachers.

Once teachers recognize that Lesson Study allows them to solve their problems through collaboration, they are excited to share their findings with their colleagues. In other words, the teachers themselves become producers of knowledge and ideas for improving their everyday teaching and learning, rather than simply using others' ideas. They can study research findings and curriculum resources to seek effective implementation of their school curriculum to maximize student learning.

We need more educators' voices on Lesson Study

Once Lesson Study becomes a method for teachers to investigate effective teaching and learning, teachers and teacher leaders are eager to share their findings. This

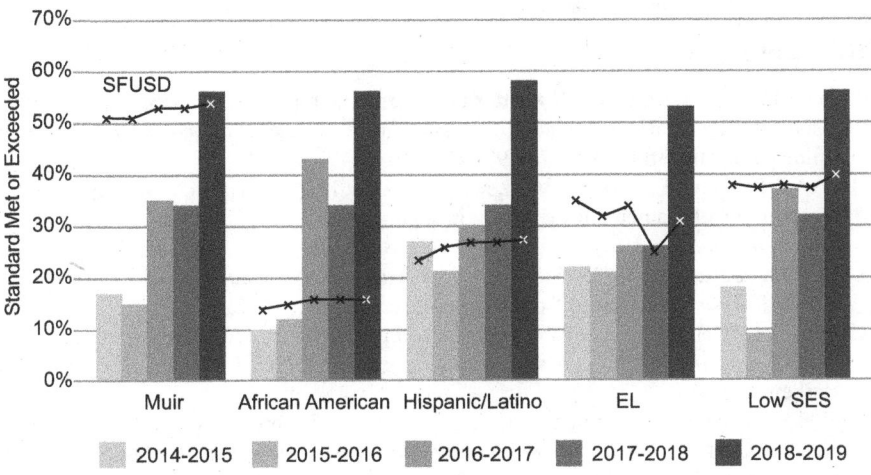

Figure 6.1 The impact of school-wide Lesson Study

Table 6.1 Contrasting views of professional development

Traditional	Lesson Study
• Begins with answer	• Begins with question
• Driven by outside "expert"	• Driven by participants
• Communication flow: trainer to teachers	• Communication flow: among teachers
• Hierarchical relations between trainer and teachers	• Reciprocal relations among learners
• Research informs practice	• Practice is research

Note: Lesson Study goes beyond traditional professional development programs. Table is based on ideas by Liptak, cited in Lewis, 2002b, p. 12. Reprinted with permission.

book was also made possible thanks to teachers who are passionate about sharing what they have learned from conducting Lesson Study at their schools. Some of these ideas may work in similar schools worldwide. They may also inspire other teachers and schools to come up with their own innovative approaches.

Many researchers who help teachers and schools conduct Lesson Study have published research articles and book chapters (e.g., Hart et al., 2011; Quaresma et al., 2018; Huang, Takahashi, & da Ponte, 2019; Takahashi, 2021). Because more and more educators have begun to experience Lesson Study and produce new ideas, we need more opportunities for teachers to share their voices. For example, this could be Lesson Study newsletters, published research lessons, or annual reports on a school's Lesson Study practice. I would like to conclude this book by envisioning a future in which other similar books will be written by teachers, teacher leaders, and school administrators who care about student learning. It is my hope that their ideas learned from Lesson Study will be published and made available worldwide.

References

Akiba, M., & Howard, C. (2021). After the race to the top: State and district capacity to sustain professional development innovation in Florida. *Educational Policy*. Advance online publication. doi:10.1177/08959048211015619

Becker, J. P., Silver, E. A., Kantowski, M. G., Travers, K. J., & Wilson, J. W. (1990). Some observations of mathematics teaching in Japanese elementary and junior high schools. *Arithmetic Teacher, 38*(2), 12–21.

Chang, L., & Rusicka, J. (1985). *Second international mathematics study United States technical report I*. Retrieved from Urbana-Champaign.

Curcio, F. (2002). *A user's guide to Japanese Lesson Study: Ideas for improving mathematics teaching* [Video and accompanying guide]. National Council of Teachers of Mathematics.

Fujii, T. (2014). Implementing Japanese Lesson Study in foreign countries: Misconceptions revealed. *Mathematics Teacher Education and Development, 16*(1), 65–83. www.merga.net.au/ojs/index.php/mted/article/view/206

Hart, L. C., Alston, A., & Murata, A. (Eds.). (2011). *Lesson Study research and practice in mathematics education: Learning* together. Springer.

Hirabayashi, I. (2002). Lesson as a drama and lesson as another form of thesis presentation. In H. Bass, Z. P. Usiskin, & G. Burrill (Eds.), *National Research Council, Studying classroom teaching as a medium for professional development: Proceedings of a U.S.-Japan workshop* (Mathematical Sciences Education Board, Division of Behavioral and Social Sciences and Education, and U.S. National Commission on Mathematics Instruction, International Organizations Board). National Academy Press.

Huang, R., Takahashi, A., & da Ponte, J. (Eds.). (2019). *Theory and practice of Lesson Study in mathematics: An international perspective*. Springer International Publishing.

Lewis, C. (2000, April 28). *Lesson Study: The core of Japanese professional development*. American Educational Research Association Meetings. https://files.eric.ed.gov/fulltext/ED444972.pdf

Lewis, C. (2002a). Does Lesson Study have a future in the United States? *Journal of the Nagoya University Department of Education,* (1), 1–23.

Lewis, C. (2002b). *Lesson Study: A handbook of teacher-led instructional change*. Research for Better Schools, Inc.

Lewis, C. (2016). How does lesson study improve mathematics instruction? *ZDM, 48*(4), 571–580. doi:10.1007/s11858-016-0792-x

Lewis, C., & Perry, R. (2014). Lesson Study with mathematical resources: A sustainable model for locally-led teacher professional learning. *Mathematics Teacher Education and Development, 16*(1), 22–42. www.merga.net.au/ojs/index.php/mted/article/view/205

Lewis, C., Perry, R., Hurd, J., & O'Connell, M. P. (2006). Lesson Study comes of age in North America. *Phi Delta Kappan, 88*(4), 273–281.

Lewis, C., & Tsuchida, I. (1998). A lesson is like a swiftly flowing river. *American Educator*, 12–51.

Mills College Lesson Study Group. (2000). *Can you lift 100 kilograms?* [DVD]. Author. https://lessonresearch.net/content-resource/can-you-lift-100-kilograms/

Mills College Lesson Study Group. (n.d.). *Schoolwide Lesson Study: John Muir Elementary School*. https://lessonresearch.net/john-muir-elementary/overview

Mullis, I. V. S., Martin, M. O., Beaton, A. E., Gonzalez, E. J., Kelly, D. L., & Smith, T. A. (1997). *Mathematics achievement in the primary school years: IEA's Third International Mathematics and Science Study (TIMSS)*. Boston College.

National Research Council. (2001). *Adding it up: Helping children learn mathematics*. National Academy Press.

National Research Council. (2002). Studying classroom teaching as a medium for professional development: Proceedings of a U.S.-Japan workshop. In H. Bass, Z. P. Usiskin, & G. Burrill (Eds.), *Mathematical Sciences Education Board, Division of Behavioral and Social Sciences and Education, and U.S. National Commission on Mathematics Instruction, International Organizations Board*. National Academy Press.

Perry, R., & Lewis, C. (2010). Building demand for research through lesson study. In C. E. Coburn & M. K. Stein (Eds.), *Research and practice in education: Building alliances, bridging the divide* (pp. 131–145). Rowman & Littlefield Publishers, Inc.

Quaresma, M., Winsløw, C., Clivaz, S., da Ponte, J. P., Ní Shúilleabháin, A., & Takahashi, A. (Eds.). (2018). *Mathematics Lesson Study around the world: Theoretical and methodological issues*. Springer International Publishing.

Stevenson, H., & Stigler, J. (1992). *The learning gap: Why our schools are failing and what we can learn from Japanese and Chinese education*. Summit.

Stigler, J., & Hiebert, J. (1999). *The teaching gap: Best ideas from the world's teachers for improving education in the classroom*. Free Press.

Stigler, J. W., Gonzales, P., Kawanaka, T., Knoll, S., & Serrano, A. (1999). The TIMSS videotape classroom study: Methods and findings from and exploratory research project on eighth-grade mathematics instruction in Germany, Japan, and the United States. *Educational Statistics Quarterly, 1*(2), 109–112. http://nces.ed.gov/pubs99/1999074.pdf

Takahashi, A. (2011). Jumping into Lesson Study – Inservice mathematics teacher education. In L. C. Hart, A. Alston, & A. Murata (Eds.), *Lesson Study research and practice in mathematics education* (pp. 79–82). Springer.

Takahashi, A. (2021). *Teaching mathematics through problem-solving: A pedagogical approach from Japan*. Routledge.

Takahashi, A., Lewis, C., & Perry, R. (2013). A US lesson study network to spread teaching through problem solving. *International Journal for Lesson and Learning Studies, 2*(3), 237–255.

Takahashi, A., & McDougal, T. (2016). Collaborative lesson research: Maximizing the impact of lesson study. *ZDM Mathematics Education, 48*(4), 513–526. doi:10.1007/s11858-015-0752-x

Yoshida, M. (1999). *Lesson study: A case study of a Japanese approach to improving instruction through school-based teacher development* (Doctoral Dissertation). University of Chicago.

Index

Page numbers in italics indicate a figure on the corresponding page.